Iberia

North ar

NORTH AND NORTH WEST SPAIN

IG 1	Planning diagram of coast	2
IG 2	Lights, fog signals and waypoints	4
IG 3	Spanish glossary	6
IG 4	Radio navigational aids	7
IG 5	Weather sources	8
IG 6	Communications	9
IG 7	Passage information	10
IG 8	Special notes for Spain	12
IG 9	Distance table (Biscay/N Spain)	12
IG 10	San Sebastián Pasajes	13
IG 11	Guetaria Motrico, Ondárroa, Lequeitio Elanchove, Bermeo	14
IG 12	Bilbao Ría de Santoña	15
IG 13	Castro Urdiales	16
IG 14	Santander San Vicente de la Barquera, Ribadesella, Avilés, Cudillero, Luarca	16
IG 15	Gijon	18
IG 16	Ría de Ribadeo	18
IG 17	Ría de Cedeira Ría de Vivero, Ría del Barquero	19
IG 18	Ría de El Ferrol Ría de Ares, Ría de Betanzos	19
IG 19	La Coruña	20
IG 20	Ría de Camariñas & Mugia Finisterre VTS	20
IG 21	Ría de Muros	20
IG 22	Ría de Arosa	22
IG 23	Ría de Pontevedra	23
IG 24	Ría de Vigo	24
IG 25	Bayona	25

PORTUGAL, SW SPAIN AND GIBRALTAR

IG 26	Planning diagram of coast	26
IG 27	Lights, fog signals and waypoints	28
IG 28	Portuguese glossary	29
IG 29	Radio navigational aids	31
IG 30	Weather sources	31
IG 31	Communications	32
IG 32	Passage information	34
IG 33	Special notes for Portugal	35
IG 34	Distance table (Portugal/SW Spain)	35
IG 35	Viana do Castelo Póvoa de Varzim	36
IG 36	Leixões (Pôrto)	36
IG 37	Figueira da Foz	37
IG 38	Nazaré	37
IG 39	Peniche	38
IG 40	Lisboa, tidal predictions/curves Cascais, Sesimbra, Setúbal	39
IG 41	Sines	44
IG 42	Lagos	44
IG 43	Portimão	45
IG 44	Vilamoura Faro/Olhão, Vila Real de Santo Antonio/Ayamonte	45
IG 45	Isla Cristina	46
IG 46	Mazagón Rio de las Pedros, Punta Umbria	47
IG 47	Chipiona Rio Guadalquivir	47
IG 48	Sevilla	48
IG 49	Bay of Cádiz: Rota, Puerto Sherry, P'to Santa Maria, P'to America	48
IG 50	Sancti-Petri	50
IG 51	Barbate	50
IG 52	Strait of Gibraltar Tarifa, Ceuta	54
IG 53	Algeciras	55
IG 54	Gibraltar, tidal predictions/curves	56
Index		Inside back cover

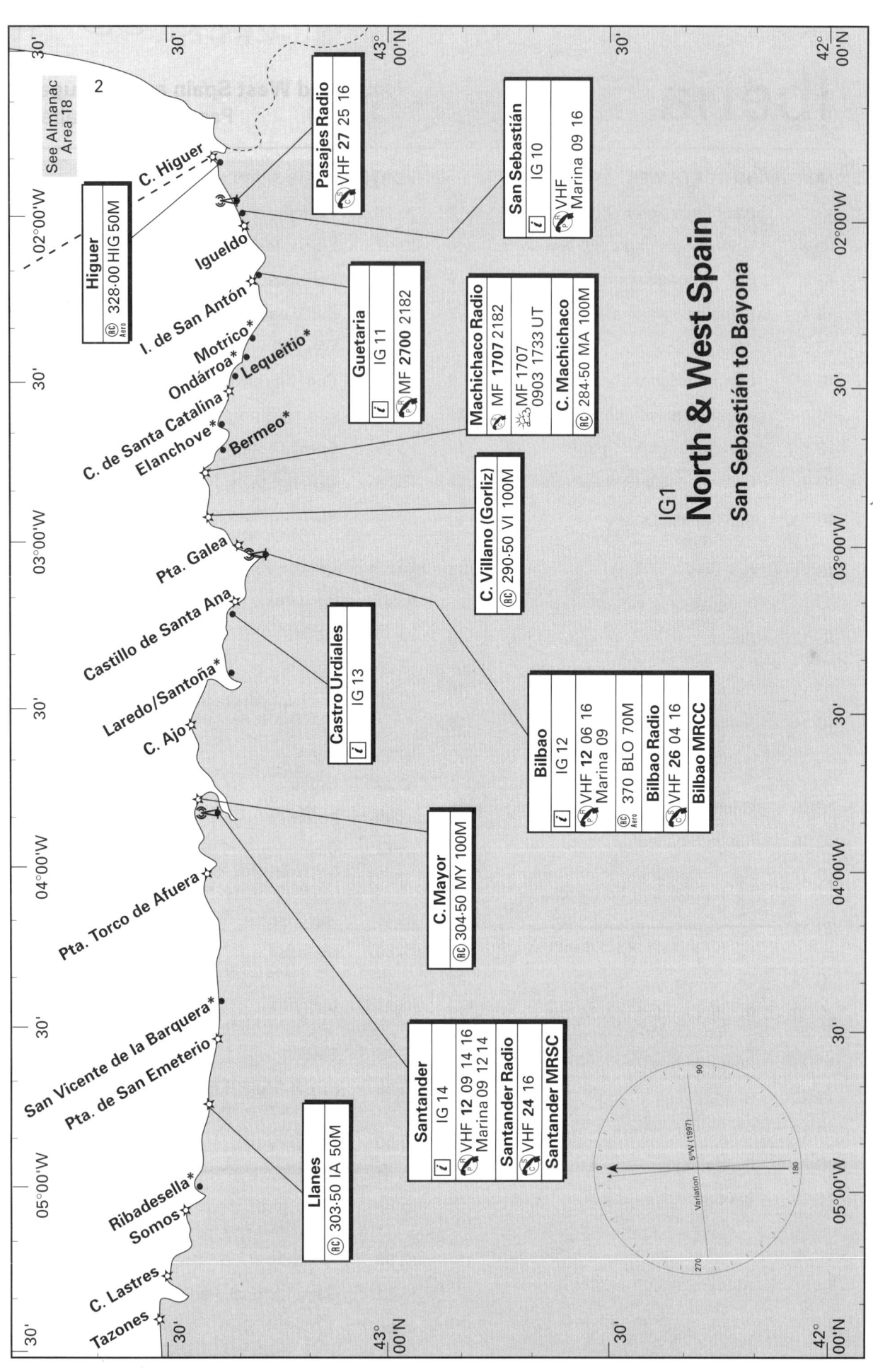

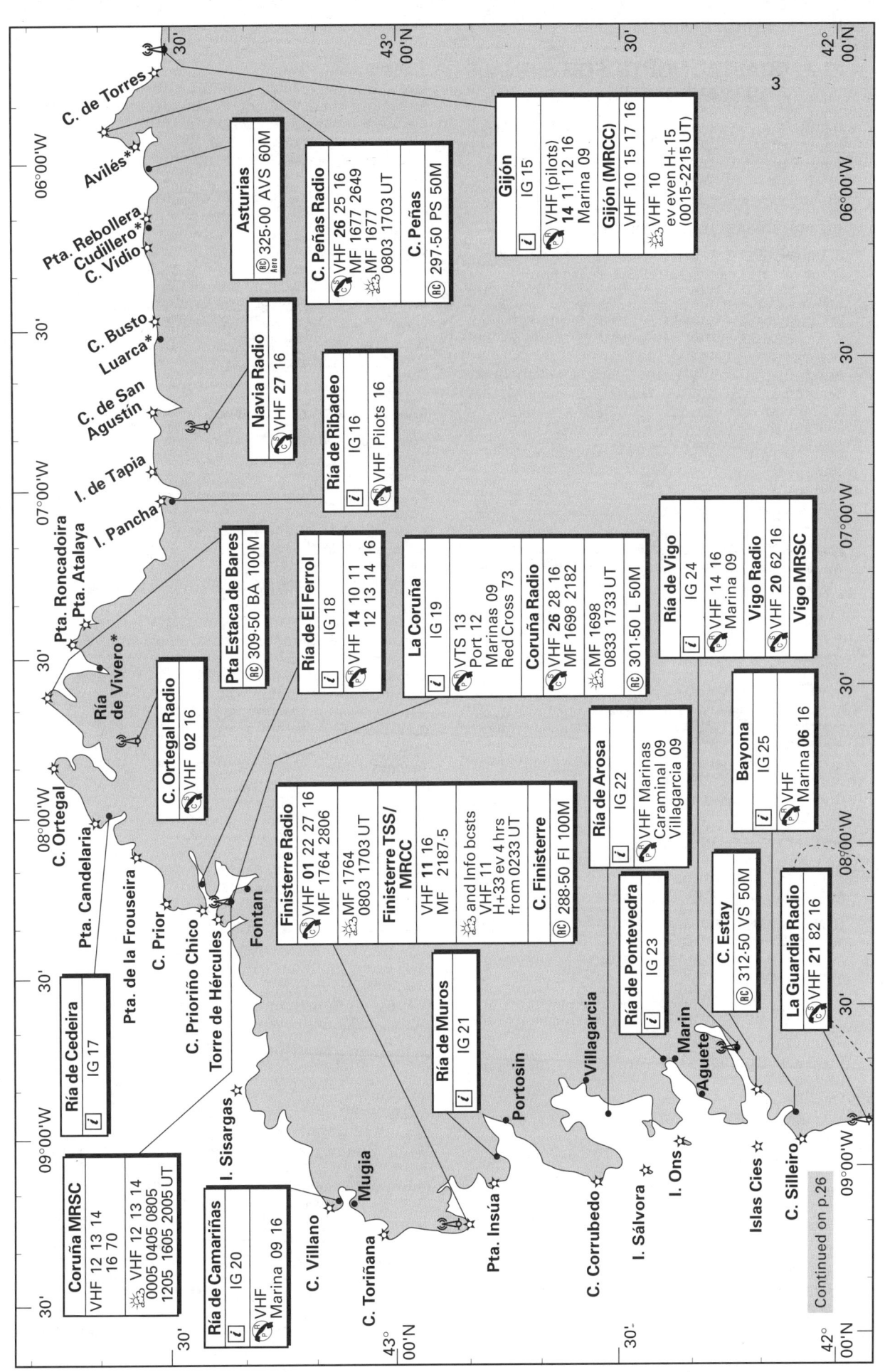

IG 2 COASTAL LIGHTS, FOG SIGNALS AND WAYPOINTS

Lights with a nominal range of 15M or more are in **bold** print, places and features in CAPITALS, and light-vessels, light floats and Lanbys in *CAPITAL ITALICS*. Fog signals are in *italics*. Useful waypoints are underlined. Generally, Spanish waypoints are referenced to ED 50.

FRENCH BORDER TO SANTANDER

C. Higuer 43°23'·59N 01°47'·44W Fl (2) 10s 63m **23M**.
Fuenterrabia, W training wall Fl (3) G 9s 9m 5M.

- PASAJES

Fairway buoy 43°21'·16N 01°56'·12W L Fl 10s; SWM.
C. La Plata 43°20'·14N 01°55'·96W Oc 4s 151m 13M; W bldg; vis 285°-250°; Racon (K).
Arando-Grande 43°20'·22N 01°55'·59W Fl R 5s 10m 11M.
Senocozulúa lts in line 154·8°: **Front** 43°19'·95N 01°55'·52W Q 67m **18M**; **rear**, 45m from front, Oc 3s 86m 18M.
SenOcozulúa dir lt Oc (2) WRG 12s 50m W6M, R3M, G3M; W twr; vis G129·5°-154·5°, W154·5°-157°, R157°-190°.

- SAN SEBASTIAN

La Concha ldg lts 158°: Front 43°18'·97N 01°59'·38W Fl R 1·5s 10m 8M; Gy mast; vis 143°-173° (intens on ldg line); rear, 25m from front, Iso R 6s 17m 8M; vis 154°-162°.
Igueldo 43°19'·43N 02°00'·55W Fl (2+1) 15s 132m **26M**.
Isla de Santa Clara 43°19'·38N 01°59'·83W Fl 5s 51m 9M.

- GUETARIA

I. de San Antón 43°18'·68N 02°12'·01W Fl (4) 15s 91m **21M**.
Shelter mole hd 43°18'·31N 02°11'·81W FG 11m 3M.

Zumaya 43°18'·20N 02°15'·00W Oc (1+3) 12s 39m 12M.

- MOTRICO/ONDÁRROA/LEQUEITIO

Malecón de Poniente hd 43°18'·50N 02°22'·90W FG 10m 2M.
Ondárroa NE bkwtr hd 43°19'·59N 02°24'·86W Fl (3) G 8s 13m 12M; Racon (G); *Siren (3) 20s*.
Lequeitio, Rompeolas de Amandarri hd 43°22'·07N 02°29'·87W Fl G 4s 10m 5M; Gy twr.
C. de Santa Catalina 43°22'·75N 02°30'·50W Fl (1+3) 20s 44m **17M**; Gy ○ twr; *Horn Mo (L) 20s*.

- ELANCHOVE/BERMEO

Elanchove Digue S hd 43°24'·30N 02°38'·19W F WR 7m W8M, R5M; vis R108°-204°, W204°-232°.
Digue Rompeolas hd 43°25'·42N 02°42'·53W Fl G 4·5s 16m 4M.
C. Machichaco 43°27'·40N 02°45'·10W Fl 7s 120m **24M**; twr; RC; *Siren (2) 60s*.
Platform Gaviota 43°30'·10N 02°41'·40W Mo (U) 10s 25m 5M; *Horn (3) 30s*.
C. Villano (Gorliz) 43°26'·08N 02°56'·62W Fl (1+2) 16s 163m **22M**; 8 sided twr.

- BILBAO

Pta Galea 43°22'·40N 03°02'·04W Fl (3) 8s 82m **19M**; twr, R&W cupola; vis 011°-227°; *Siren Mo (G) 30s*.
Pta Galea bkwtr hd, 43°22'·84N 03°04'·59W, Fl R 6s 19m 6M.
Pta Lucero bkwtr hd 43°22'·74N 03°04'·95W Fl G 4s 21m 14M; Racon (X).
Santurce W bkwtr hd 43°20'·86N 03°01'·81W Fl (2) G 12s 11m 4M. (Port authority bldg).
Contramuelle de Algorta hd 43°20'·60N 03°01'·56W, Fl (4) R 14s 18m 6M, W stone twr.
Dir lt, 43°20'·40N 03°00'·70W, Q 22m 11M, W tr on house, vis 119°-135°

- CASTRO URDIALES

Castillo de Santa Ana 43°23'·13N 03°12'·81W Fl (4) 24s 46m **20M**; W ▲ twr; *Siren Mo (C) 60s*.
Rompeolas N hd 43°22'·92N 03°12'·47W Fl G 3s 12m 6M.

- LAREDO/SANTOÑA

Laredo N bkwtr hd 43°24'·96'N 03°25'·12W FR 9m 2M.
Santoña ldg lts 283·5°: Front, 43°26'·40N 03°27'·52W Fl 2s 6m 8M; ▼ on Gy twr; rear,0·75M from front, Oc (2) 5s 13m 11M; ○ on twr; vis 279·5°-287·5°.
Pta Pescador 43°27'·90N 03°26'·05W Fl (3+1) 15s 38m 9M; Gy ○ twr.
C. Ajo 43°30'·80N 03°35'·20W Oc (3) 16s 69m **17M**; ○ twr.

- SANTANDER

C. Mayor 43°29'·48N 03°47'·37W Fl (2) 10s 89m **21M**; W ○ twr; *Horn (2) 40s*; RC.
I. Mouro 43°28'·47N 03°45'·27W Fl (1+2) 21s 37m 11M.
Puntal ldg lts 236°: Front Pta Rabiosa 43°27'·58N 03°46'·35W Q 7m 6M; rear, 100m from front, Iso R 4s 10m 6M.
No. 3 buoy 43°27'·79N 03°46'·13W Fl (3) G 9s; SHM.
Dársena de Molnedo, mole E hd 43°27'·75N 03°47'·39W QG 10m 3M.

SANTANDER TO CABO PEÑAS

Pta Torco de Afuera 43°26'·58N 04°02'·52W Fl (1+2) 24s 33m **22M**; W twr; obscured close inshore 091°-113°.
Suances ldg lts 146°: Front 43°26'·27N 04°01'·99W Q 8m 5M; rear, Punta Marzán 210m from front, Iso 4s 12m 5M.

- SAN VICENTE DE LA BARQUERA

Pta de la Silla 43°23'·68N 04°23'·44W Oc 3·5s 42m 13M; twr; vis 115°-250°; *Horn Mo (V) 30s*.
I. Peña Menor 43°23'·80N 04°23'·02W FlWG 2s W7M, G6M; G twr; vis G175°-235°, W235°-045°.
Pta San Emeterio 43°23'·90N 04°32'·10W Fl 5s 66m **20M**.
Llanes, Pta de San Antón 43°25'·20N 04°44'·90W Oc (4) 15s 16m **15M**; W 8-sided twr; RC.
Somos 43°28'·43N 05°04'·88W Fl (1+2) 12s 113m **21M**; twr.
Ribadesella Pta del Caballo 43°28'·15N 05°03'·89W Fl (2) R 6s 10m 5M; ○ twr; vis 278·4°-212·9°.
C. Lastres 43°32'·20N 05°17'·90W Fl 12s 116m **23M**. W ○ tr.
Lastres bkwtr hd 43°30'·96N 05°15'·81W Fl (3) G 9s 11m 4M.
Tazones 43°32'·80N 05°24'·00W Oc (3) 15s 125m **15M**; W 8-sided twr; *Horn Mo (V) 30s*.

- GIJÓN

Banco Las Amasucas S buoy 43°34'·60N 05°39'·70W Q (6) + L Fl 15s; SCM.
C. de Torres 43°34'·38N 05°41'·87W Fl (2) 10s 80m **18M**.
Dique Principe de Asturias 43°34'·32N 05°40'·49W Fl G 3s 22m 9M; G twr.

Candás Punta del Cuerno 43°35'·71N 05°45'·56W Oc (2)10s 38m 13M; R twr, W ho; *Horn Mo (C) 60s*.
Luanco ldg lts 255°: Front, mole hd, 43°36'·98N 05°47'·27W Fl R 3s 4m 4M; rear, 240m from front, Oc R 8s 8m 4M.
C. Peñas 43°39'·42N 05°50'·78W Fl (3) 15s 115m **35M**; Gy 8 sided twr; RC; *Siren Mo (P) 60s*.

CABO PEÑAS TO PUNTA DE LA ESTACA DE BARES

Ría de Avilés buoy 43°35'·80N 05°57'·60W Fl G 5s; SHM.
Avilés 43°35'·80N 05°56'·63W Oc WR 5s 38m **20/17M**; W □ twr; vis R091·5°-113°, W113°-091·5°; *Siren Mo (A) 30s*.
Ría de Avilés ent chan S side 43°35'·64N 05°56'·41W Fl (2) 7s 10m 5M; vis 106°-280°; W ○ twr, G band.
Puerto de S. Estaban W bkwtr elbow 43°34'·13N 06°04'·66W Fl (2) 12s 19m **15M**; W **I**, B bands: *Horn Mo (N) 30s*.
San Esteban de Pravia ldg lts 182·2°: Front 43°33'·90N 06°04'·58W FR 6m 3M; rear, 160m from front, FR 10m 3M.
Pta Rebollera 43°34'·00N 06°08'·50W Oc (4) 16s 42m **16M**; W 8-sided twr; *Siren Mo (D) 30s*.
C. Vidio 43°35'·60N 06°14'·70W Fl 5s 99m **25M**; ○ twr; *Siren Mo (V) 60s*.
C. Busto 43°34'·25N 06°28'·10W Fl (4) 20s 84m **21M**.

Coastal Lights, Fog Signals and Waypoints

- LUARCA
Pta Focicón (Blanca) 43°33'·03N 06°31'·85W Oc (3) 15s 63m 14M; W □ twr; Siren Mo (L) 30s.
Ldg lts 170°: Front 43°32'·82N 06°32'·02W Fl 5s 18m 2M; rear, 41m from front, Oc 4s 25m 2M.
Dique del Canouco hd 43°32'·97N 06°32'·04W Fl (3) R 9s 22m 5M; ○ Tr.

Ría de Navia outfall buoy 43°34'·25N 06°43'·55W Fl Y 10s; SPM.
C. de San Agustín 43°33'·90N 06°43'·98W Oc (2) 12s 70m **18M**; W ○ twr, B bands.
I. de Tapia 43°34'·50N 06°56'·60W Fl (1+2) 19s 22m **18M**.

- RÍA DE RIBADEO
I. Pancha 43°33'·47N 07°02'·43W Fl (3+1) 20s 26m **21M**; W ○ twr, B bands; Siren Mo (R) 30s.
Pta de la Cruz 43°33'·47N 07°01'·65W Fl (2) 7s 16m 5M.
Ldg lts 140°: Front, Pta Aerojo, 43°32'·90N 07°01'·43W QR 18m 5M; rear, 228m from front, Oc R 4s 24m 5M. R ♦, W twrs.
Ldg lts 205°: Front, Muelle de García, 43°32'·56N 07°02'·16W V QR 8m 3M; rear, 178m from front, Oc R 2s 18m 3M.

Ría de Foz trg wall hd 43°34'·50N 07°14'·60W Fl G 3s 3m 10M.
Piedra Burela 43°39'·80N 07°20'·80W Q (3) 10s 11m 7M; ECM. *bo de San Ciriaco*
Pta Atalaya 43°42'·10N 07°26'·11W Fl (5) 20s 39m **20M**.
Alúmina Port N bkwtr hd 43°43'·04N 07°27'·48W Fl (2) WG 8s 18m 4M; vis W110°-180°, G180°-110°.
Pta Roncadoira 43°44'·05N 07°31'·50W Fl 7·5s 92m **21M**.

- RÍA DE VIVERO/RÍA DEL BARQUERO
Pta Socastro 43°43'·15N 07°36'·32W Fl G 7s 18m 5M; W twr.
Pta de Faro 43°42'·81N 07°34'·93W Fl (2) R 14s 18m 5M.
Cillero dique hd 43°40'·99N 07°36'·05W FR 9m 4M.
Isla Coelleira 43°45'·58N 07°37'·67W Fl (4) 24s 87m 8M.
Pta del Castro 43°44'·54N 07°40'·39W Q (2) R 6s 14m 5M.
Pta de la Barra 43°44'·58N 07°41'·21W Fl WRG 3s 15m 5M; vis 213°-240°, R240°-255°, G255°-213°; W ▲ twr.
Vicedo N. Pier head, Fl(2) R 7s 9m 3M; 43°44'·38N 7°40'·42W
S. Pier head, Fl(2) G 6s 9m 3M; 43°44'·37N 7°40'·42W

PUNTA DE LA ESTACA DE BARES TO CABO VILLANO

Pta de la Estaca de Bares 43°47'·30N 07°41'·00W Fl (2) 7·5s 99m **23M**; 8 sided twr & house; Siren Mo (B) 60s; RC.
Espasante outer bkwtr 43°43'·2N 07°48'·90W Fl G 5s 6m 3M.
Cariño bkwtr hd 43°44'·12N 07°51'·63W Fl G 2s 12m 3M.
C. Ortegal 43°46'·30N 07°52'·20W Oc 8s 122m **18M**; W ○ twr, R band.
Pta Candelaria 43°42'·70N 08°02'·80W Fl (3+1) 24s 87m **21M**.
Punta de Sarridal Oc WR 6s 59m 11M; vis R shore-145°, W145°-172°, R172°-shore; 43°39'·72N 08°04'·64W

- RÍA DE CEDEIRA
Piedra de Media Mar 43°39'·45N 08°04'·71W Fl (2) 5s 12m 4M; W ○ twr.
Pta Promontorio 43°39'·12N 08°04'·12W Oc (4) 10s 24m 11M.
Dique de Abrigo hd 43°39'·37N 08°04'·11W Fl (4) R 13s 10m 3M.

Pta de la Frouseira 43°37'·10N 08°11'·25W Fl (5) 15s 73m **20M**.
C. Prior 43°34'·10N 08°18'·70W Fl (1+2) 15s 105m **22M**; 6-sided twr; vis: 055·5°-310°; ~~Siren Mo (P) 25s~~.

- RÍA DE EL FERROL
C. Prioriño Chico 43°27'·58N 08°20'·31W Fl 5s 34m **23M**.
Lt buoy 43°27'·49N 08°18'·72W Fl G 2s; SHM.
Batería de S.Cristobal 43°27'·99N 08°18'·17W Oc (2) WR 10s.
Ldg lts 085·4°: Front Pta de San Martín Fl 1·5s 10m 3M; rear, 704m from front Oc 4s 3M; both W trs.
No. 3 buoy 43°27'·88N 08°16'·38W Fl (2) G 9s; SHM.
Dársena de Curuxeiras mole hd 43°28'·60N 08°14'·57W Fl (4) R 11s 6m 3M; W I.
Sada. Dique de Abrigo hd 43°21'·83N 08°14'·44W Fl (3) G 9s 9m 4M; G twr.

- LA CORUÑA
Torre de Hércules 43°23'·23N 08°24'·31W Fl (4) 20s 104m **23M**; □ twr; RC; Siren Mo (L) 30s.
Banco Yacentes buoy 43°24'·75N 08°22'·92W Fl (5) Y 20s; SPM.
Ldg lts 108·5°: Front, Pta Mera 43°23'·07N 08°21'·17W Oc WR 4s 54m 8M; vis R000°-023°, R100·5°-105·5°, W105·5°-114·5°, R114·5°-153°; Racon (M); rear, 300m from front, Fl 4s 79m 8M; vis 357·5°-177·5°; both W twrs.
Ldg lts 182°: Front, Pta Fiaiteira 43°20'·66N 08°22'·16W Iso WRG 2s 28m W4M, R3M, G3M; vis G146·4°-180°, W180°-184°, R184°-217·6°; rear, 380m from front, Oc R 4s 53m 3M; ○ twrs.
Dique d'Abrigo hd 43°21'·97N 08°22'·38W Fl G 3s 16m 6M.

Malpica shelter mole hd 43°19'·40N 08°48'·20W Fl G 3s 19m 4M.
Islas Sisargas 43°21'·60N 08°50'·60W Fl (3) 15s 108m **23M**; W cupola on twr, on W bldg; Siren (3) 30s.
Pta del Roncudo 43°16'·58N 08°59'·37W Fl 6s 36m 10M.
Pta Lage 43°14'·00N 09°00'·60W Fl (5) 20s 64m **20M**. W twr.
Corme mole hd 43°15'·78N 08°57'·79W Fl (2) R 5s 13m 3M R ○ tr.
Puerto de Lage dique N hd 43°13'·42N 08°59'·85W Fl G 3s 16m 4M; twr.
C. Villano 43°09'·60N 09°12'·70W Fl (2) 15s 102m **28M** Y/Gy 8-sided twr; RC; Racon (M); Siren Mo (V) 60s.
Punta Nariga 43°19'·3N 08°54'·51W Fl (3+1) 20s 53m 22M; W ○ Ltho

CABO VILLANO TO PORTUGUESE BORDER

- RÍA DE CAMARIÑAS/MUGIA
Ldg lts 079·7°: Front, Pta Villueira 43°07'·45N 09°11'·47W Fl 5s 14m 9M; W ○ twr, R ♦; rear, 610m from front, Pta del Castillo Iso 4s 25m 11M; W ○ twr, R bands; vis 078·2°-081·2°.
Pta de Lago 43°06'·68N 09°09'·92W Oc (2) WRG 6s 15m W6M, R4M, G4M; vis W029·5°-093°, G093°-107·8°, W107·8°-109·1°, R109·1°-139·3°, W139·3°-213·5°; ○ twr.
Camariñas outer bkwtr hd 43°07'·58N 09°10'·62W FR 1M.
FV pier hd 43°07'·71N 09°10'·83W Fl (3) R 8s 9m 4M.
Pta de la Barca 43°06'·86N 09°13'·09W Oc 4s 13m 6M.
Mugia bkwtr hd 43°06'·42N 09°12'·67W Fl (2) G 10s 12m 3M.

C. Toriñana 43°03'·27N 09°17'·74W Fl (2+1) 15s 63m **24M**.
C. Finisterre 42°53'·00N 09°16'·22W Fl 5s 141m **23M**; 8-sided twr; RC; Racon (O); obscd when brg >149°.
Finisterre bkwtr hd 42°54'·63N 09°15'·28W Fl R 2s 13m 4M.
I. Lobeira Grande 42°52'·92N 09°11'·02W Fl (3) 15s 16m 9M.
Carrumeiro Chico 42°54'·43N 09°10'·66W Fl (2) 7s 6m 6M; △ twr, R band; IDM.
C. Cée 45°55'·06N 09°10'·92W Fl (5) 13s 25m 7M; Gy twr.
Corcubión bkwtr hd 42°56'·77N 09°11'·29W Fl (2) R 8s 9m 4M.

Puertocubelo bkwtr hd 42°48'·48N 09°08'·03W Fl G 2s 9m 4M.
Pta Insúa 42°46'·36N 09°07'·47W F WR 26m **W15M**, R14M and Oc (1+2) WR 20s W15M, R14M; vis FR 308°-012·5°, OcR 012·5°-044·5°, Oc W 044·5°-093°; FW 093°-172·5° (see AC 1756).

- RÍA DE MUROS
Pta Queixal 42°44'·43N 09°04'·65W Fl (2+1) 12s 25m 9M.
C. Reburdiño 42°46'·28N 09°02'·81W Fl (2) R 6s 16m 7M.
Muros outer mole hd 42°46'·72N 09°03'·22W Fl (4) R 13s 8m 4M; W I.
El Freijo mole hd 42°47'·68N 08°56'·52W Fl (2) R 5s 7m 4M.
Portosin bkwtr hd 42°45'·96N 08°56'·80W Fl (2) G 5s 7m 3M.
Pta Cabeiro 42°44'·48N 08°59'·30W Fl WRG 4s 35m W9M, R6M, G6M; W ○ twr; vis R 054·4°-058·5°, G058·5°-099·5°, W099·5°-189·5°.
El Son bkwtr hd 42°43'·80N 08°59'·97W Fl G 5s 4m 6M.
Pta Sofocho 42°41'·98N 09°01'·61W Fl 5s 27m 4M; ○ twr.

C. Corrubedo 42°34'·67N 09°05'·30W clear sector 089·4°- about 200° Fl (3+2) R 20s 30m **15M**; and danger sector about 332°-089·4°, Fl (3) R 20s 30m **15M**; Gy ○ twr Siren (3) 60s.
Corrubedo bkwtr hd 42°34'·41N 09°04'·10W Iso WRG 3s

10m W6M, R4M, G3M; ○ twr; vis R000°-016°, G016°-352°, W352°-000°.

- **RÍA DE AROSA**

I Sálvora 42°27'·93N 09°00'·71W Fl (3+1) 20s 38m **21M**; clear sector 217°-126°. Fl (3) 20s dangerous sector 126°-160°.
Piedras del Sargo E rk 42°30'·38N 09°00'·41W QG 11m 8M: W ▲ twr, G band.
Puerto de Aguiño bkwtr hd 42°31'·17N 09°00'·80W FR 12m 3M.
Santa Eugenia bkwtr hd 42°33'·68N 08°59'·12W Fl (2) R 7s 8m 4M; W ○ twr.
Isla Rúa 42°33'·02N 08°56'·27W Fl (2+1) WR 21s 24m 13M.
Puebla del Caramiñal E bkwtr hd 42°36'·35N 08°55'·77W Oc G 2s 8m 4M; ○ twr.
Rianjo outer bkwtr hd 42°39'·13N 08°49'·35W Fl G 3s 9m 5M.
Villagarcia de Arosa Muelle del Ramal hd 42°36'·20N 08°46'·25W Iso 2s 2m 10M; ○ twr.
Isla de Arosa, St Julian N bay mole hd 42°34'·04N 08°52'·08W FR 7m1M; R masonry l.
Pta Caballo 42°34'·40N 08°52'·95W Fl (4) 11s 11m 10M.
San Martin del Grove mole hd 42°29'·92N 08°51'·40W, Oc G 2·5s 9m 3M; B & W chequered ○ twr.
Bajo Pombeiriño lt bn 42°28'·95N 08°56'·70W Fl (2) G 12s 13m 8M; G ▲ twr, G band.
Islas de Sagres 42°30'·59N 09°02'·84W Fl 5s 23?m 8M

- **RÍA DE PONTEVEDRA**

Isla Ons 42°23'·01N 08°56'·06W Fl (4) 24s 126m **25M**; twr.
Playa del Curro, pier 42°22'·68N 08°55'·77W 7m 2M R ○ tr.
Bajo Camouco 42°23'·84N 08°54'·65W Fl (3) R 18s 10m 8M.
Bajo Picamillo 42°24'·37N 08°53'·37W Fl G 5s 10m 8M; twr.
Porto Novo mole hd 42°23'·70N 08°49'·14W Fl (3) R 6s 8m 4M.
Sangenjo mole hd 42°23'·94N 08°48'·20W Fl (4) R 12s 8m 4M.
Cabezo de Morrazan buoy 42°22'·51N 08°46'·95W Iso R 5s; PHM.
Rajo mole hd 42°24'·1N 08°45'·3W Fl (2) R 8s 9m 3M R ○ tr.
Combarro mole hd 42°25'·82N 08°42'·21W Fl (2) R 8s 7m 3M.
I. Tambo 42°24'·56N 08°42'·38W Oc (3) 8s 33m 11M; W○twr.
Marin bkwtr hd 42°24'·01N 08°42'·21W Fl (3) G 7s 8m 6M.
Bueu N mole hd 42°19'·86N 08°47'·05W Fl G 3s 9m 4M G tr.
Bajo Mourisca 42°20'·96N 08°49'·02W Fl (2) G 7s 10m 5M.
Aldan mole hd 42°16'·8N 08°49'·4W Fl (2) R 10s 7m 5M.
Pta Couso 42°18'·69N 08°51'·21W Fl (3) WG 9s 18m 10/8M.

- **ISLAS CÍES**

Monte Agudo 42°14'·65N 08°54'·10W Fl G 5s 24m 9M; W twr.
No 2 buoy (Roca Omear) 42°14'·65N 08°51'·80W Fl (4) R 10s; PHM; *Bell*.
Monte Faro 42°12'·94N 08°54'·78W Fl (2) 8s 185m **22M**.
Pta Canabal 42°12'·81N 08°54'·63W Fl (2) 20s 63m 9M; W twr.
C. Vicos 42°11'·58N 08°53'·38W Fl (3) R 9s 92m 7M; W twr.
Islote Boiero 42°10'·82N 08°54'·48W Fl (2) R 8s 22m 6M.

- **RÍA DE VIGO**

C. Home ldg lts 129°: Front 42°15'·23N 08°52'·28W Fl 3s 36m 9M; W twr; rear Pta Subrido, 815m from front, Oc 6s 52m 11M; W twr; both vis 090°-180°.
Cangas outer mole hd 42°15'·70N 08°46'·74W Fl (1+2) R 12s 8m 3M; R ○ twr.
Pta Areiño. **La Guia** 42°15'·65N 08°42'·04W Oc (2+1) 20s 35m **15M**; R ○ twr.
C. Estay ldg lts 069·3°: **Front**, 42°11'·20N 08°48'·73W Iso 2s 16m **18M**; *Horn Mo (V) 60s*; Racon (B), **rear**, 660m from front, Oc 4s 48m **18M**; both R ▲ twrs. W hands; vis 066·3° 072·3°.
Pta Lameda 42°09'·47N 08°50'·90W Fl (2) G 8s 27m 5M; W tr.

- **BAYONA**

Las Serralleiras 42°08'·87N 08°52'·57W Fl G 4s 10m 6M.
Ldg lts 084°: Front Cabezo de San Juan 42°08'·26N 08°50'·08W Q (2) 4s 7m 6M; W ▲ twr; vis 081·5°-084·5°; rear Panjón, 1·05M from front, Oc R 4s 17m 9M; W ▲ twr.
C. Silleiro 42°06'·34N 08°53'·70W Fl (2+1) 15s 83m **24M**; RC.
La Guardia 41°54'·10N 08°52'·80W Fl R 5s 11m 5M; R ▲ twr.

GLOSARIO ESPAÑOL IG 3

See IG 28 for Portuguese glossary

A. NAVIGATION	NAVEGACIÓN
Beacon (Bn)	baliza
Buoy	boya
Can (PHM buoy)	cilíndrica
Cone, conical (SHM buoy)	cónica
Isolated danger (IDM buoy)	peligro aislado
Landfall (SWM buoy)	atterragio
Special mark (SPM buoy)	marca especiá
Leading line, transit	enfilación
Port (side)	babor
Starboard	estribor
Topmark	marca de tope
Red, (R)	rojo
White (W)	blanco
Black (B)	negro
Yellow (Y)	amarillo
Green (G)	verde
Stripe/band	francas verticales/horiz
Lighthouse	faro
Lightship	buque faro
Fixed (F)	luz fija
Alternating (Al)	luz alternativa
Flashing	luz destellos
Quick flashing (Q)	luz centelleante
Occulting (Oc)	luz de Ocultaciones
Leading light	luz de enfilación
Obscured	Oculto
Whistle	silbato
Bell	campana
North (N)	norte
South (S)	sur
East (E)	este
West (W)	oeste
Chart Datum (CD)	bajamar escorada
High Water (HW)	pleamar (PM)
Low Water (LW)	bajamar (BM)
Neaps (np)	aguas muertas
Springs (sp)	aguas vivas
Slack water, stand	repunte
Range	repunte
Tide tables	Tabla de mareas
Tidal stream atlas	Atlas de mareas
Flood/ebb stream	entrante/vaciante
Rate/set (tide)	velocidad/dirrección
Knots (kn)	nudos
Height, headroom, clearance	altura
Draught	calado
Bar	barra
Bay	bahia, ensenada
Sandhill, dunes	dunas, médano
River	río
Strait(s)	estrecho
Point, headland	punta
Island	isla
Bridge	puente
Conspicuous (conspic)	conspicuo
Rock	roca, piedra
Wreck	naufragio
Shoal	restinga, bajo

Reef	arrecife
Anchorage	fondeadero
Breakwater, mole	dique, estacada
Basin	dársena, doca

B. FACILITIES / FACILIDADOS

Yacht harbour, marina	dársena de yates
Harbour Master	capitanía
Coastguard (CG)	guardacostas
Customs (#)	aduana
Registration number	matricula
Length overall (LoA)	eslora total
Beam	manga
Draught	calado
Mooring buoy	boya de amarre
Dredged	dragado
Jetty	muelle
Slipway (slip)	varadero
Lock	esclusa
Lifeboat (LB)	bota salvavidas
Chandlery (CH)	apetrachamento
Crane (C)	grua
Boatyard (BY)	astillero
Sailmaker (SM)	velero
Engineer (ME)	ingeniero
Boat hoist (BH)	portico elevador
Fresh water (FW)	agua potable
Diesel (D)	gas-oil, gasoleo, diesel
Petrol (P)	gasolina
Paraffin	parafina
Methylated spirits	alcohol desnaturalizado

C. METEOROLOGY / METEOROLOGIA

High (anticyclone)	anticiclón
Ridge (high)	dorsal
Low (depression)	depresión
Pressure, rise/fall	presión, subida/caida
Front, warm/cold	frente, cálido/frío
Calm (F0)	calma
Light airs (F1)	ventolina
Light breeze (F2)	flojito
Gentle breeze (F3)	flojo
Moderate breeze (F4)	bonancible
Fresh breeze (F5)	fresquito
Strong breeze (F6)	fresco
Near gale (F7)	frescachón
Gale (F8)	duro
Severe gale (F9)	muy duro
Storm (F10)	temporal
Squall	turbonada
Shower	aguacero
Rain	lluvia
Hail	granizada
Thunderstorm	tempestad
Cloudy	nubloso
Mist	neblina
Fog	niebla
Swell	mar de leva
Shortosteep (sea state)	mar corta
Rough sea	mar gruesa
Slight sea	marejadilla
overfalls (tide race)	escarceos
Breakers	rompientes

Radio navigational aids 7

IG 4 RADIO NAVIGATIONAL AIDS

4.1 GPS

The only radio navigation aid able to provide continuous fixing coverage is GPS. ~~Differential GPS is planned for installation at C. Finisterre, operating on 289·00 kHz with a range of 60M.~~ *DGPS beacons are on trial at: P. Estaca de Bares 43°47'·17N 07°41'·09W 310·0 ǂ — Cabo Finisterre 42°53'·00N 09°16·23W 289·0 ǂ 60*

4.2 Loran

Loran-C (Lessay chain GRI 6731) should give about 100m accuracy along the North Spain coast, decreasing to about 470m at the limit of ground wave coverage at 40°N.

4.3 Racons (Radar beacons)

NORTH & NORTHWEST SPAIN

Puerto de Pasajes	43°20'·17N 01°55'·39W	K
Ondárroa NE bkwtr hd	43°19'·60N 02°24'·86W	G
Bilbao (Pta Lucero bkwtr)	43°22'·73N 03°04'·96W	X
Pta Mera (La Coruña)	43°23'·08N 08°21'·17W	M
C. Villano	43°09'·68N 09°12'·60W	M
C. Toriñana	43°01'·82N 09°16'·43W	T
C. Finisterre	42°53'·00N 09°16'·23W	O
~~C. Estay (Ría de Vigo)~~	~~42°11'·20N 08°48'·73W~~	~~B~~
C. Corrubedo	*42°34'·67N 09°05'·30W*	*K*
Isla Rúa	*42°33'·02N 08°56·27W*	*W*

4.4 RDF Beacons (marine and aeronautical)

NORTH & NORTHWEST SPAIN

San Sebastián Aero			
43°23'·25N 01°47'·65W	328·00	**HIG**	50M
C. Machichaco			
43°27'·45 02°45'·08W	284·50	**MA**	100M
Bilbao Aero			
43°19'·43N 02°58'·43W	370·00	**BLO**	70M
C. Mayor			
43°29'·48N 03°47'·37W	304·50	**MY**	100M
Llanes			
43°25'·20N 04°44'·90W	303·50	**IA**	50M
C. Peñas			
43°39'·42N 05°50'·80W	297·50	**PS**	50M
Asturias Aero			
43°33'·57N 06°01'·50W	325·00	**AVS**	60M
P. Estaca de Bares			
43°47'·17N 07°41'·07W	309·50	**BA**	100M
Torre de Hércules			
43°23'·23N 08°24'·30W	301·50	**L**	50M
C. Villano			
43°09'·68N 09°12'·60W	290·50	**VI**	100M
C. Finisterre*			
42°53'·00N 09°16'·23W	288·50	**FI**	100M
C. Estay			
42°11'·19N 08°48'·73W	312·50	**VS**	50M
C. Silleiro			
42°06'·33N 08°53'·70W	293·50	**RO**	100M

*Differential GPS is planned for C. Finisterre, operating on 289·00 kHz with a range of 60M.

IG 5 WEATHER SOURCES

5.1 General
Daily forecasts can usually be obtained from YCs or Hr Mrs. The geographic limits of the weather forecast areas used by UK, French, Spanish and Portuguese authorities do not necessarily coincide, even though similar names are used. Forecasts issued by different nations may also differ markedly in scope and accuracy.

Spanish forecast areas not shown on Fig. 1 below are:
Area 1 Gran Sol (S. Ireland to Dover Strait and S to 48°N)
Area 2 Vizcaya (48°N to the N boundaries of Cantabrico and Finisterre forecast areas).
Area 5 Azores (W of Areas 4 and 6, out to 30°W).

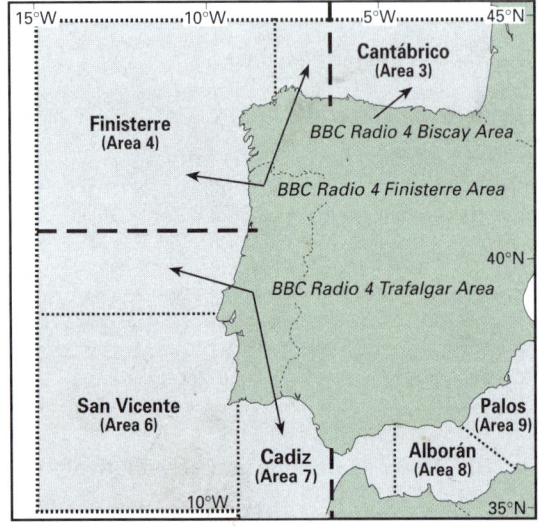

Fig. 1 Spanish forecast areas
.... Spanish Forecast Areas
— — — UK Forecast Areas (BBC)

5.2 BBC Shipping Forecasts
BBC Radio 4 shipping forecasts for Sea Areas Biscay, Finisterre and Trafalgar (see Fig. 1) are broadcast daily on 198 kHz (1515m) at 0048, 0555, 1355 and 1754 (LT)* and can normally be received in N and NW Spain, especially at night. *~~Times are likely to change wef April 1998~~. Sea Area Trafalgar is included only in the 0048 forecast.

5.4 Radio facsimile ~~forecast~~
Stations transmitting MF/HF SSB radio facsimile weather charts covering Iberia and adjacent sea areas include ~~Madrid~~ ~~(ECA7)~~ and Rota (US Navy) (AOK). Their schedules and frequencies are detailed in Section 30.5 of this Guide.

5.3 Navtex
Navtex is described in the Almanac, Chapter 6, 6.4.25. Stations which can be received across the Bay of Biscay and in N and NW Spain, together with identification letter, times (UT) of navigational warnings and forecasts (**shown in bold type**), all in English, are:

Corsen (A) 0000 0400 0800 **1200** 1600 2000
    Weather for W Brittany, N & S Biscay, Small Sole, Galicia and Romeo
Coruña (D) 0030 0430 0830 **1230** 1630 2030
    Weather for Spanish areas 1-6

5.5 Recorded telephone forecasts
Within North and North West Spain call ☎ 906 365 372 for Offshore areas 1-4, ie Gran Sol, Vizcaya (N Biscay), Cantábrico and Finisterre. Also forecasts for Coastal waters from the French to the Portuguese borders. All in Spanish. The service is only available within Spain and to Autolink-equipped vessels.

5.6 Broadcasts of Gale Warnings and Forecasts for Offshore & Coastal Waters
Gale warnings and forecasts are broadcast, as shown below, by the Spanish National Radio, Coast Radio Stations and by MRCC/MRSCs.

Radio Nacional de España (National Radio) broadcasts in Spanish storm warnings, synopsis and 12h or 18h forecasts for Spanish Areas 3 & 4 at 1100, 1400, 1800 and 2200LT. Stations and frequencies are:
San Sebastián 774 kHz; **Bilbao** 639 kHz; **Santander** 855 kHz; **Oviedo** 729 kHz; **La Coruña** 639 kHz.

Coast Radio Stations (see also 6.1)
The following CRS broadcast in Spanish gale warnings, synopsis and forecasts for Areas 1-5 at the times and on the MF frequencies shown (there are no VHF weather broadcasts by CRS):
Machichaco at 0903 1733UT on 1707 kHz; **Cabo Peñas** at 0803 1703UT on 1677 kHz; **La Coruña** at 0833 1733UT on 1698 kHz; **Finisterre** at 0803 1703UT on 1764 kHz.

Coastguard MRCC/MRSC (see also 6.2)
broadcast in Spanish (in English) gale warnings on receipt; synopsis and forecasts for the Areas shown at the times (UT) and VHF channels listed below:

Bilbao MRCC every 4h from 0033 on VHF Ch 10 for Areas 2-4
Santander MRSC at every 4h from 0245 on VHF Ch 11 for Areas 2-4
Gijón MRCC at every even H+15 from 0015 to 2215 on VHF Ch 10 16 for Areas 3 & 4.
Coruña MRSC every 4h from 0005 on VHF Ch 12, 13 14 for Areas 1-5.
Finisterre MRCC every 4h from 0233 on VHF Ch 11 for Areas 1-5. 0633 1033 1433 1833 2233 (UT)
Vigo MRSC at every 4h from 0015 on VHF Ch 10 for Areas 3-6.

IG 6 COMMUNICATIONS

6.1 COAST RADIO STATIONS
Initial call should be made on Ch 16 (H24) using the callsign of the remotely controlled station which will switch you to a working channel. Dedicated Autolink channels are shown in italics; before calling verify that the Autolink chan is not in use. Traffic lists are only broadcast on MF, not VHF, every odd H+33, 0333-2333UT, except 2133.

NORTH SPAIN
Stations are remotely controlled from Bilbao Comms Centre.

Pasajes Radio　　　　　43°17'N 01°55'W　　　　VHF **27** *25 (Autolink)*
　Navigation warnings (in Spanish): Ch 27 on receipt, after next silence period and at 0803 1503.

Machichaco Radio　　　43°27'N 02°45'W　　　　No VHF
　MF: Transmits **1707**, 2182 kHz (H24); receives on 2132, 2045, 2048, 2182 (H24). *Traffic lists:* 1707 kHz. *Navigation warnings:* 1707 kHz. Urgent warnings on receipt, after next silence period and at 0033 0433 0833* 1233 1633 2033*. *Other warnings in **English**/Spanish.

BILBAO RADIO　　　　43°22'N 03°02'W　　　　VHF **26** *04 (Autolink)*
　Navigation warnings (in Spanish): Ch 26 on receipt, after next silence period and at 0933 1533.

Santander Radio　　　　43°25'N 03°36'W　　　　VHF **24** *24 (Autolink)*
　Navigation warnings (in Spanish): Ch 24 on receipt, after next silence period and at 0803 1503.

Cabo Peñas Radio　　　43°39'N 05°51'W　　　　VHF **26** *25 (Autolink)*
　MF: Transmits **1677**, 2182 kHz (H24); receives on 2102, 2045, 2048, 2182 (H24). *Autolink 2649, 3231.* *Traffic lists:* 1677 kHz. *Navigation warnings (in Spanish):* Ch 26 on receipt, after next silence period and at 0903 1603. Other warnings at 0833 2033.

Navia Radio　　　　　　43°25'N 06°50'W　　　　VHF **27**
　Navigation warnings (in Spanish): Ch 27 on receipt, after next silence period and at 0833 1533.

NORTH WEST SPAIN
Stations are remotely controlled from Coruña Comms Centre.

Cabo Ortegal Radio　　　43°35'N 07°47'W　　　　VHF **02**
　Navigation warnings (in Spanish): Ch 02 on receipt, after next silence period and at 0903 1603.

CORUÑA RADIO　　　　43°22'N 08°27'W　　　　VHF **26** *28 (Autolink)*
　MF: Transmits **1698**, 2182 kHz (H24); receives on 2123, 2045, 2048, 2182 (H24). *Traffic lists:* 1698 kHz. *Navigation warnings (in Spanish):* Ch 26 on receipt, after next silence period and at 0803 1503. 1698 kHz: Urgent warnings on receipt, after next silence period and at 0003 0403 0803* 1203 1603 and 2003*. *Other warnings in **English** & Spanish.

Finisterre Radio　　　　42°54'N 09°16'W　　　　VHF **01 22** *27 (Autolink)*
　MF: Transmits **1764**, 2182 kHz (H24); receives on 2108, 2045, 2048, 2182 (H24). *Autolink 2806, 3283.* *Traffic lists:* 1764 kHz. *Navigation warnings (in Spanish):* Ch 01 22 on receipt, after next silence period and at 0903 1603. 1764 kHz: Urgent warnings on receipt, after next silence period and at 0033 0433 0833* 1233 1633 2033*. *Other warnings in **English** & Spanish.

Vigo Radio　　　　　　42°10'N 08°41'W　　　　VHF **20** *62 (Autolink)*
　Navigation warnings (in Spanish): Ch 20 on receipt, after next silence period and at 0803 1503.

La Guardia Radio　　　41°53'N 08°52'W　　　　VHF **21** 82
　Navigation warnings (in Spanish): Ch 21 on receipt, after next silence period and at 0903 1603.

6.2 COASTGUARD STATIONS

MRCC/MRSC

MRCCs and MRSCs primarily handle Distress, Safety and Urgency communications. Madrid MRCC coordinates SAR on the N coast of Spain through six MRCC/MRSC. All stations monitor VHF Ch 16 and 2182 kHz H24. Digital Selective Calling (DSC) is operational or planned as shown below; dates may change considerably, so consult Notices to Mariners for latest information. Stations also broadcast weather messages as shown in IG 5.6, and navigational warnings as shown below (all times are UT). They do not handle link calls.

Station	Position	Phone/Fax	Channels
BILBAO MRCC MMSI 002240996	43°21'N 03°02'W	☎ 94 483 9286; 📠 94 483 9161	Ch 10: Nav warnings every 4 hours from 0233 DSC Ch 70, 2187·5 kHz
Santander MRSC MMSI 002241009	43°28'N 03°43'W	☎ 942 213030; 📠 942 213638	Ch 11: Nav warnings every 4 hours from 0045 DSC Ch 70, ~~2187·5 kHz (both planned 1997)~~
GIJÓN MRCC MMSI 002240997	43°37'N 05°42'W	☎ 985 326 050; 📠 985 320 908 Call: *Gijón Traffic*	Ch 10, 15, 17, 16 (H24), 2182, 2657 kHz (H24) Ch 10: Nav warnings every H+15 DSC Ch 70, 2187·5 kHz
Coruña MRSC MMSI 002240992	43°22'N 08°23'W	☎ 981 209 548; 📠 981 209 518 Call: *Coruña Traffic*	Ch 12, 13, 14, 2657 kHz (H24) Ch 13: Nav warnings every 4 hours from 0205 DSC Ch 70, 2187·5 kHz
FINISTERRE MRCC MMSI 002240993	42°42'N 08°59'W	☎ 981 767 320; 📠 981 767 740 Call: *Finisterre Traffic*	Ch 11. Ch 11: Nav warnings every 4 hours from 0033 DSC Ch 70, 2187·5 kHz
Vigo MRSC MMSI 002240998	42°10'N 08°41'W	☎ 986 297403; 📠 986 290455 Call: *Vigo Traffic*	Ch 10: Nav warnings every 4 hours from 0215 DSC Ch 70, 2187·5 kHz (both planned 1997)

PASSAGE INFORMATION   IG 7

BIBLIOGRAPHY

The *South Biscay Pilot* (Adlard Coles, 4th edition 1995) covers from the Gironde to La Coruña. The *Atlantic Spain and Portugal* guide (Imray/RCC, 3rd edition 1995) covers from El Ferrol to Gibraltar. The *Bay of Biscay Pilot* (Admiralty, NP 22) covers from Pte de Penmarc'h to Cabo Ortegal, whence The *W coasts of Spain and Portugal Pilot* (Admiralty, NP 67) continues south to Gibraltar. *Portos de Galicia* (English/Spanish) has good photographs. *Guia del Navegante*, also in English, has fair cover of SW Spain, but is limited elsewhere. For a Spanish glossary, see IG 3.

BAY OF BISCAY (S): WIND, WEATHER AND SEA

Despite its reputation, the S part of the Bay is often warm and settled in summer when the Azores high and Spanish heat low are the dominant weather features. NE'lies prevail in sea area Finisterre in summer and gales may occur twice monthly, although forecast more frequently. Atlantic lows can bring W'ly spells at any time. SE or S winds are rare, but wind direction and speed often vary from day to day. Sea and land breezes can be well developed in the summer. Off N Spain *Galernas* are dangerous squally NW winds which blow with little warning. Coastal winds are intensified by the Cordillera Cantábrica (2615m mountains). Rainfall is moderate, increasing in the SE, where thunder is more frequent. Sea fog occurs May-Oct, but is less common in winter.

No tidal stream atlases are published; streams are weak offshore, but can be strong in narrow channels and around headlands. Surface current much depends on wind: in summer it sets SE ½-¾kn, towards the SE corner of B of Biscay, thence W along N coast of Spain. When crossing the Bay, allow for some set to the E, particularly after strong W winds.

CROSSING THE BAY OF BISCAY (chart 1104)

From the English Channel to Cabo Villano, the track (210°/365M) from Ushant lies close to busy shipping lanes, but a track (213°/355M) from, say, Raz de Sein to Cabo Villano is offset from the shipping route. From Scilly or Eire the direct track lies in about 7°-8°W. Within the Bay itself, sailing down the French coast is attractive and allows a 200M passage from, say, La Rochelle to Santander. S of the Gironde a missile range may inhibit coastal passage. The Continental Shelf where depths plummet from 150m to over 4000m in only 30M, can cause dangerous seas in swell and bad weather. It trends SE from about 60M SW of Ile de Sein and is clearly depicted on AC 1104. The British Decca chain 1B may cover to about 100M S of Ushant by day, but there are no Decca chains in France or Spain. Loran-C (chain 6731) covers N Spain well and extends down to about 40°N (Figueira da Foz (IG 37)) and westward to 15°W.

The Atlantic swell, rarely experienced in UK waters except off the W coasts of Ireland and Scotland, runs in mainly from W or NW. It is not a problem well offshore, except over the Continental Shelf in bad weather. Considerable rolling is probable and other yachts may be lost to view in the troughs. Closer inshore swell will exacerbate the sea state and render entry to, and exit from, lee shore hbrs dangerous. For example, with a 2m swell running, crossing a bar with 4m depth over it and breaking seas would be foolhardy. In winter some hbrs are closed for weeks at a time, more particularly along the Portuguese coast.

FRENCH BORDER TO CABO ORTEGAL (charts 1102, 1105, 1108)

The N coast of Spain is bold and rocky. In clear visibility the peaks of the Cordillera Cantábrica may be seen from well offshore. Although the coast is mostly steep-to, it is best to keep 2-3M offshore (beyond the 50m contour) to avoid short, steep seas breaking on isolated shoals. Many major lights are sited so high as to be obscured by low cloud/mist in onshore weather. Of the many small rivers, most are obstructed by bars and none are navigable far inland.

Between the French border and Santander are several interesting fishing hbrs (Motrico, Lequeitio, Elanchove and Laredo/Santoña) offering anchorage or AB. Marinas are still the exception rather than the rule. Pasajes is a port of refuge, but also a large, commercial/fishing port. 3M to the W, San Sebastián (IG 10) is an attractive ⚓ but exposed to the NW. Bilbao (IG 12) and Santander (IG 14) are major cities and ferry ports for the UK, with marinas. In NW gales ports of refuge for yachts are Guetaria (IG 11), Bermeo, Bilbao, Castro Urdiales (IG 13) and Santander.

Beyond Santander hbrs are increasingly far apart. Gijón (IG 15) with a modern marina and San Ciprian offer refuge only within their vast commercial ports. Cudillero and Luarca are small fishing hbrs worth visiting, but beware of swell particularly at Cudillero. Ría de Ribadeo (IG 16) at 7°W is the first of the rías altas (upper or northern), sunken estuaries not unlike a Scottish sea loch. The Ría de Vivero, Ría del Barquero and the Ensenada de Santa Marta, to E and W of Pta de la Estaca de Bares, offer many attractive ⚓s sheltered from all but N/NE winds; but beware S/SW winds off the mountains being accelerated by funnelling effects through the valleys.

CABO ORTEGAL TO CABO FINISTERRE (charts 1111, 3633)

Cabo Ortegal should be rounded at least 2M off due to the offlying needle rks, Los Aguillones. Most of the major headlands should be given a good offing to avoid fluky winds and, in some cases, pinnacle rks. The deeply indented coast begins to trend WSW, with 600m high mountains rising only 5M inland. Here also many major lights are sited so high as to be obscured by low cloud/mist in onshore weather. Tidal streams set SW on the ebb and NE on the flood. Any current tends to run SW, then S.

The Ría de Cedeira (IG 17) is entered 3M SSW of Pta Candelaria lt ho. This small attractively wooded ría offers refuge by day/night to yachts unable to round Cabo Ortegal in strong NE'lies. Several banks along this stretch break in heavy weather when they should be passed well to seaward. About 20M further SW, having passed Pta de la Frouseira, Cabo Prior, Pta del Castro and C. Prioriño Chico, all lit, is the ent to Ría de El Ferrol (IG 18), a well sheltered commercial and naval port. La Coruña (IG 19), 5M SSW, can be entered in all weathers and has far better yacht facilities. Between these two large ports are the quiet and little frequented rías de Ares and Betanzos, the latter with a marina at Fontan/Sada. W of the conspic Torre de Hércules lt ho a wide bight is foul as far as Islas Sisargas, three islets 2M offshore. At Pta Nariga (43°19'·3N 08°54'·6W) is a new, but unlit lt ho. 8M SW, between Pta del Roncudo and Pta de Lage, is the attractive Ría de Corme y Lage.

Cabo Villano, a rky headland with lt ho and conspic wind generators, is a possible landfall after crossing Biscay. A pinnacle rk awash at CD, lurks 4ca NW of it. The ent to Ría de Camariñas (IG 20), the last of the rías altas and a safe refuge, is close S. 20M W/NW of Cabo Villano and Finisterre is a TSS; see IG 20 for Maritime Traffic Service. Cabo Toriñana and Cabo Finisterre both have charted dangers, mostly pinnacle rks, lying up to 1M offshore; hence the popular name *Costa del Morte* for this wild and magnificent coast.

CABO FINISTERRE TO RIO MIÑO (chart 3633)

NE of Cabo Finisterre the Ensenada del Sardineiro and Ría de Corcubión are sheltered ⚓s except in S'lies. From 5 to 11M south, beware various islets, reefs and shoals N and W of Pta Insua, notably Las Arrosas and Bajo de los Meixidos; the latter can be passed inshore if making for Ría de Muros (IG 21).

The 40M stretch to Cabo Silleiro embraces an impressive cruising ground in its own right, namely the four major rías bajas (lower or southern), from N to S: Muros, Arosa (IG 22), Pontevedra (IG 23) and Vigo (IG 24). The last three rías are sheltered from onshore winds by coastal islands at their mouth, but in summer NE winds can blow strongly down all rías, usually without raising any significant sea. Arosa, the largest ría, runs 12M inland and is up to 6M wide; Villagarcia is the pricipal hbr and marina. All resemble large and scenic Scottish sea lochs with fishing hbrs, many sheltered anchorages and interesting pilotage. Ría de Vigo is notable for the beautiful Islas Cies, the port of Vigo and the pleasant passage hbr of Bayona (IG 25).

The Río Miño (*Minho* in Portuguese), 15M S of Cabo Silleiro, forms the N border between Spain and Portugal. The river ent is difficult and best not attempted.

For Passage Information along the coasts of Portugal and Andalucía see IG 32.

SPECIAL NOTES FOR SPAIN   IG 8

Regions/Provinces: Spain is divided into 17 autonomous regions, ie in this Guide, the Basque Country, Cantabria, Asturias, Galicia and Andalucía. Most of the regions are sub-divided into provinces, eg in Galicia: Lugo, La Coruña, Pontevedra and Orense (inland). The province is shown below the name of each main port.

Language: This Guide recognises the significant regional differences (eg Basque, Gallego), but in the interests of standardisation uses the spelling of Castilian Spanish where practicable.

Charts: Spanish charts (SC) are obtainable from Instituto Hidrográfico de la Marina, Tolosa Latour 1, 11007 Cádiz, ☎ (956) 599412, 📠 275358; order from Seccion Economica by post/bank transfer or in person for cash. Also from Centro Nacional de Informacion Geografica, 6 Calle Lopez Doriga, Santander. Hbrs are often charted at larger scale than British Admiralty charts.

Time: Spain keeps UT+1 as standard time; DST (Daylight Saving Time) is kept from the last Sunday in March until the Saturday before the 4th Sunday in October, as in other EU nations. Note: Portugal keeps UT as standard time and UT+1 as DST.

Telephone: To call Spain from UK dial 00-34, then the area code, *preceded by* less the initial 9, followed by the ☎ number. To call UK from Spain, dial 07-44, then area code, less the initial 0, then the ☎ number.

Emergencies: ☎ 900 202 202 for Fire, Police and Ambulance. *Rioja Cruz* (Red Cross) operate LBs.

Public Holidays: Jan 1, 6; Apr 10 (Good Friday); 1 May (Labour Day); June 11 (Corpus Christi); Aug 15 (Assumption); Oct 12 (National Day); Nov 1 (All Saints Day); Dec 6, 8 (Immaculate Conception), 25.

Representation: Spanish National Tourist Office, 57 St James St, London SW1A 1LD; ☎ 0171 499 0901, 📠 0171 629 4257.
British Embassy, Calle de Fernando el Santo 16, 28010 Madrid; ☎ (91) 319 0200, 📠 319 0423. There are British Consuls at Bilbao, Santander, Vigo, Seville and Algeciras (qv).

Buoyage: IALA Region A system is used. However buoys may lack topmarks, be unpainted (or wrongly painted) more often than in N Europe. Preferred chan buoys (RGR, Fl (2+1) R and GRG, Fl (2+1) G) are quite widely used.

Documents: Spain, although an EU member, still asks to check paperwork. This can be a time-consuming, repetitive and inescapable process. The only palliatives are courtesy, patience and good humour. Organise your papers to include:
Personal – Passports; crew list, ideally on headed paper with the yacht's rubber stamp, giving DoB, passport nos, where joined/intended departure. Certificate of Competence (Yachtmaster Offshore, ICC/HOCC etc). Radio Operator's certificate. Form E111 (advised for medical treatment).
Yacht – Registration certificate, Part 1 or SSR. Proof of VAT status. Marine insurance. Ship's Radio licence. Itinerary, backed up by ship's log.

Access: Ferries sail from UK to Santander and Bilbao (10.0.4 in the Almanac). There are flights from UK to Bilbao, Santiago de Compostela and Jerez; also Sevilla, Madrid, Malaga and Gibraltar. Internal flights via Madrid to Asturias, La Coruña and Vigo. Bus services are usually good and trains adequate, except in the more remote regions.

Currency: Approx 236 pesetas/£ in Dec 1997.

TIMES (UT) OF SUNRISE & SUNSET

		JAN		FEB		MAR		APR		MAY		JUN		JUL		AUG		SEP		OCT		NOV		DEC	
		1	15	1	15	1	15	1	15	1	15	1	15	1	15	1	15	1	15	1	15	1	15	1	15
BILBAO	SR	0744	0741	0728	0710	0648	0624	0554	0530	0505	0448	0435	0431	0435	0445	0502	0517	0536	0551	0609	0625	0646	0705	0724	0737
	SS	1647	1701	1723	1742	1801	1818	1838	1854	1913	1929	1945	1953	1956	1950	1934	1915	1847	1822	1753	1729	1704	1648	1637	1637
LA CORUÑA	SR	0806	0803	0750	0732	0710	0646	0616	0551	0527	0509	0456	0452	0457	0506	0523	0538	0557	0613	0631	0647	0708	0727	0746	0759
	SS	1708	1723	1745	1804	1822	1840	1900	1916	1935	1951	2007	2016	2018	2012	1956	1937	1909	1844	1815	1751	1725	1709	1659	1658

DISTANCE TABLE: BAY OF BISCAY AND NORTH/NORTH WEST COASTS OF SPAIN   IG 9

Approximate distances in nautical miles are by the most direct route, whilst avoiding dangers and allowing for Traffic Separation Schemes. Places in *italics* are in Eire, England and West coast of France. See also Distance Table at IG 34.

1. *Crosshaven (Cork)*	**1**																			
2. *Tuskar Rk*	85	**2**																		
3. *Longships*	144	130	**3**																	
4. *Ushant (Créac'h)*	235	230	100	**4**																
5. *Bénodet*	303	298	168	68	**5**															
6. *Belle Ile*	409	336	206	106	52	**6**														
7. *Ile d' Yeu*	460	387	257	157	102	51	**7**													
8. *La Rochelle*	526	453	323	223	168	117	66	**8**												
9. *Arcachon*	515	510	380	280	227	187	138	98	**9**											
10. San Sebastián	572	567	437	337	290	246	203	177	87	**10**										
11. Cabo Machichaco	550	549	419	319	272	233	195	177	101	34	**11**									
12. Bilbao (ent)	549	550	420	320	273	238	201	187	116	50	16	**12**								
13. Santander	534	541	411	311	266	234	206	198	133	82	48	36	**13**							
14. Gijón	508	528	391	298	267	254	237	248	205	162	128	116	90	**14**						
15. Cabo Peñas	500	523	385	293	264	252	237	249	211	170	136	126	96	10	**15**					
16. Ría de Ribadeo	499	520	391	308	308	279	271	288	252	223	189	179	149	63	53	**16**				
17. Cabo Ortegal	484	510	388	307	307	290	289	313	285	258	224	214	184	98	88	40	**17**			
18. La Coruña	508	538	418	338	327	326	326	351	323	296	262	252	222	136	126	78	38	**18**		
19. Cabo Villano	523	556	439	365	356	358	359	383	355	328	294	284	254	168	158	110	70	43	**19**	
20. Bayona	594	627	510	436	427	429	430	454	426	399	365	355	325	239	229	181	141	114	71	**20**

SAN SEBASTIÁN (Donestia) IG 10

Guipúzcoa 43°19'·42N 01°59'·32W

CHARTS
AC 1181, 1102; SC 19B, 944, 945; SHOM 6558, 6786
TIDES
Standard Port PTE DE GRAVE (⟵); ML No data; Zone –0100

Times				Height (metres)			
High Water		Low Water		MHWS	MHWN	MLWN	MLWS
0000	0600	0500	1200	5·4	4·4	2·1	1·0
1200	1800	1700	2400				
Differences PASAJES							
–0050	–0030	–0015	–0045	–1·2	–1·3	–0·5	–0·5
SAN SEBASTIÁN							
–0110	–0030	–0020	–0040	–1·2	–1·2	–0·5	–0·4

SHELTER
Fair; the ⚓ becomes lively or dangerous in any NW/N'lies when heavy swell enters. Options: ⚓ S of Isla Santa Clara or SW of YC, in both cases clear of moorings and with ⚓ buoyed; or pick up a vacant ⚬. Possible AB on pontoon at ent to the over-crowded basins (no berths). The city's elegance is not matched by facilities for visiting yachts.
NAVIGATION
WPT 43°19'·85N 01°59'·90W, 338°/158° from/to narrows (between Isla de Santa Clara and Monte Urgull), 0·5M. Avoid La Bancha on which the sea breaks in swell and heavy weather. Open up the bay before standing in.
LIGHTS AND MARKS
Monte Urgull (huge statue of Virgin Mary) is prominent from all directions; from E it looks like an island. Monte Igueldo is only slightly less obvious. Isla de Santa Clara is low and inconspicuous and not seen until the nearer appr. Ldg marks (grey masts) are hard to see, but a large R & W Palladian-styled villa with prominent pilasters is conspic on 158°. Ldg lts, although intensified, may be masked by shore lts.
RADIO TELEPHONE
VHF Ch 09 (H24).
TELEPHONE (Dial code 943)
Hr Mr 351816; CG Emerg'y 900 202 202; LB 94 483 9286; Met 274030; Auto 906 365320; ⊞ 945 454000; Police 092; Brit Consul 94 415 7600.
FACILITIES
Darsena de la Concha (Yacht/FV basins, both solid with local boats). **Real Club Náutico** ☎ 423575, 📠 431365, R, Bar, Ice. **City:** Gaz, V, R, Bar, ⓑ, ✉, ⊞, ⛴ & ✈ Fuenterrabia (25km); ✈ also at Bilbao (92km) and ferry.

ADJACENT HARBOUR

PASAJES (Pasaia) 43°20'·21N 01°55'·69W (3·5M E of San Sebastián). Tides in IG 10. AC 1181; SC 9440; SHOM 6375. A port of refuge, but also a busy FV and commercial hbr with no yacht facilities. SWM buoy, L Fl 10s, (43°21'·16N 01°56'·12W) bears 340°/1·0M from the hbr ent which is a 200m wide cleft in the cliffs, marked by ☆s Fl R 5s & G 5s 18m 11M and ☆ Oc 4s 151m 13M on the W side. Tfc sigs, 3 vert lts, high on E cliff show: ⓇⓌⓇ = no entry; ⒼⓌⒼ = no exit; ⒼⓌⓇ = port closed. Enter on 155° transit of Dir lt, Oc (2) WRG 12s (vis W 154·5°-157°) with ldg lts: front Q 67m 18M, rear Oc 3s 86m 18M. Inner chan, dredged 10m, is only 100m wide, but well marked/lit; do not impede large vessels/FVs. Yachts may ⚓ on E side of fairway, in bay to NE of Dir lt; E or W of fairway if space allows, but subject to wash. Basins further S are commercial/FV. VHF Ch 09, 16. Hr Mr ☎ (943) 351816, 📠 351348.

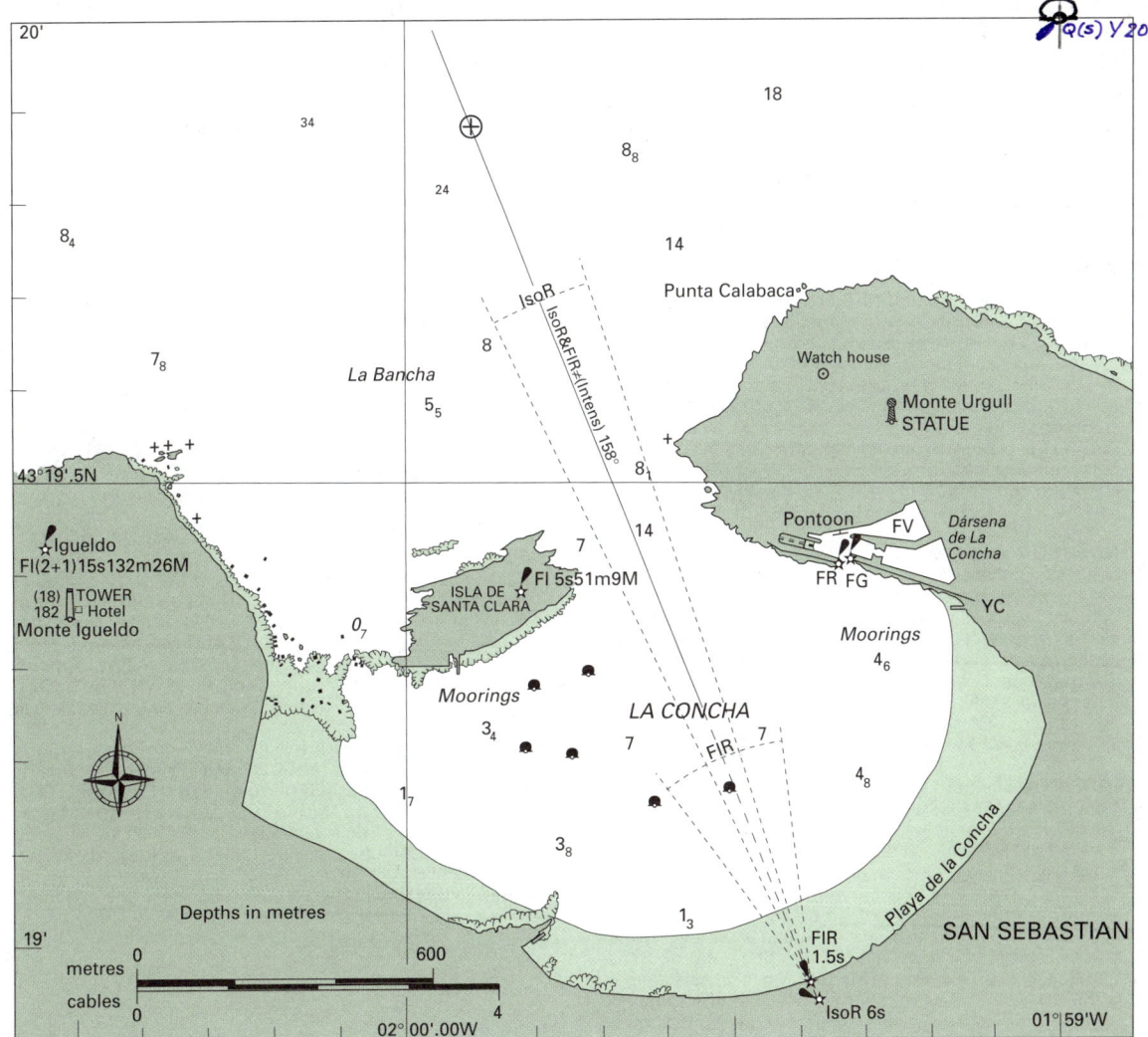

GUETARIA (Getaria)   IG 11
Guipúzcoa 43°18'·30N 02°11'·80W

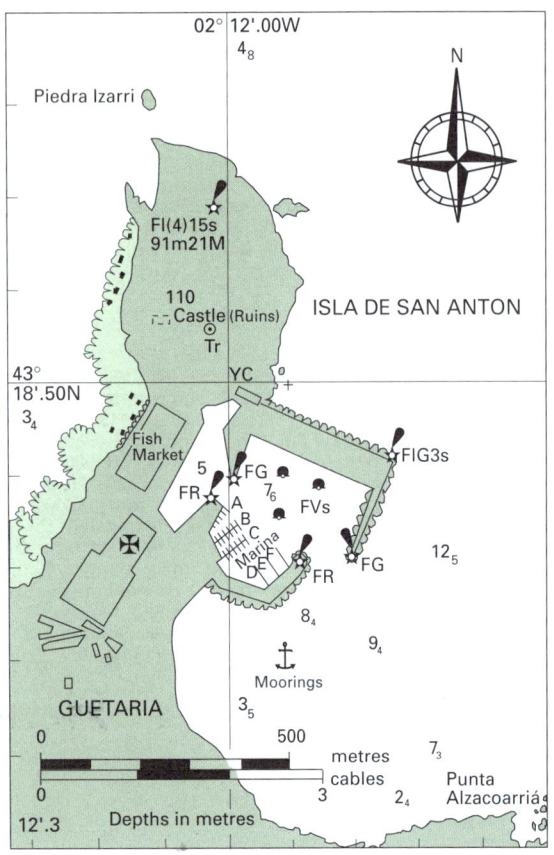

CHARTS
AC 1171, 1102; SC 303A, 944, 943, 128; SHOM 6379, 6786
TIDES
Standard Port PTE DE GRAVE (←—); ML No data; Zone –0100

Times				Height (metres)			
High Water		Low Water		MHWS	MHWN	MLWN	MLWS
0000	0600	0500	1200	5·4	4·4	2·1	1·0
1200	1800	1700	2400				
Differences GUETARIA							
–0110	–0030	–0020	–0040	–1·0	–1·0	–0·5	–0·4
LEQUEITIO							
–0115	–0035	–0025	–0045	–1·2	–1·2	–0·5	–0·4
BERMEO							
–0055	–0015	–0005	–0025	–0·8	–0·7	–0·5	–0·4

SHELTER
Good in marina (5m); no nominated ♥ berths, but try A-C pontoons. FVs fill the inner hbr and also berth on N and W sides of outer hbr, but yachts may berth here if FVs are at sea; or pick up vacant FV buoy. No ⚓ in hbr which is generally foul. In fair weather ⚓ on sand or pick up a buoy close S of hbr ent, but exposed to E'lies.
NAVIGATION
WPT 43°18'·90N 02°11'·00W, 045°/225° from/to hbr ent, 8½ca. Appr is straightforward day/night in fair visibility.
LIGHTS AND MARKS
Lts as chartlet. Isla de San Anton, from E or W is shaped like a toy mouse, but is more conical seen from N; it is topped by conspic lt ho and castle ruins. Beaches 2M SE of hbr are easily seen from N.
RADIO TELEPHONE
VHF Ch 09 H24.
TELEPHONE (Dial code 943)
Hr Mr 140413; ⌗ via Hr Mr; Met 906 365 320; Auto 906 365365; LB 900 202 202.
FACILITIES
Marina (270, inc few ♥), ☎ 140201, FW, AC, BH (30 ton);
Club Náutico y Pesca ☎ 140201, R, Bar.
Services: Slip, D, Sh, CH, SM, ME, El, C (5 ton);
Town: V, R, Bar, Gaz, ☎, ✉, ⇌; ✈ Bilbao (UK ferry).

OTHER HARBOURS BETWEEN GUETARIA AND BILBAO

MOTRICO (Mutriku) 43°18'·80N 02°22'·80W. Tides: interpolate times/hts between Guetaria (IG11) & Lequeitio. AC 1102; SC 16a; SHOM 6379. Small FV hbr (4m) in rky inlet. Easily seen from the E, but from the W open up the hbr & town before approaching; caution rky ledges off Pta de Cardal. Ldg lts/marks 236·5°: front, FR 10m 2M on S bkwtr; rear, FR 63m 2M on clock tr (hard to see by day). N bkwtr, FG 10m 2M. In summer shelter is good except with NE'ly swell/surge. ⚓ outside the N bkwtr in 5m near spending beach and slip; or ⚓/moor inside the 23m wide ent to the W; or AB on SE quay if FVs are away. Hr Mr (943) 601328. **Facilities** on FV quay: FW, AC, D, P, C (5 ton), Ice; Slip. **Town** R, V, Bar.

ONDÁRROA 43°19'·60N 02°24'·85W. Tides: interpolate between Guetaria (IG11) and Lequeitio. AC 1171; SC 707; SHOM 6379. Important fishing port, close W of Pta Saturrarán, offers good shelter, but not recommended for yachts. In strong N'lies ent is difficult and only to be attempted near HW. Approach from N-NE and keep close round the NE bkwtr. Channel, moorings and inner basin all dredged 4m. Inner basin is oily, dirty and full of FVs. Lts: NE bkwtr hd, Fl (3) G 8s 13m 12M, siren (3) 20s, Racon; S bkwtr Fl (2) R 6s 7m 6M; Inner basin FG and FR. **Facilities**: as for a fishing hbr.

LEQUEITIO (Lekeitio) 43°22'·06N 02°29'·85W. Tides: IG11. AC 1171; SC 642A, 943; SHOM 6379. Enter to W of steep-sided Isla de San Nicolas, joined to the mainland by a training wall covered at HW. ✠ dome is conspic. Pass about 5m off the outer bkwtr head on 205° to clear rky shoals and the small detached bkwtr close to the E. In the basin possible AB on W quay or S mole, ⚓ in about 3m or pick up a vacant buoy at the S end. FVs use the N end. In settled weather ⚓ further out, to E or W of Isla de San Nicolas. Lts: Cabo de Santa Catalina, Fl (1+3) 20s 44m 17M, Horn Mo (L) 20s, is conspic 0·75M to the NW. Outer bkwtr Fl G 4s 10m 5M; detached bkwtr Fl (2) R 8s 5m 4M; inner basin ent, FG & FR. Club de Pesca ☎ (94) 6840500. **Facilities**: FW, D, P, CH, Slip, ME, V.

ELANCHOVE (Elantxobe) 43°24'·30N 02°38'·18W. Tides: interpolate between Lequeitio and Bermeo. AC 1171; SC 321A, 942/3; SHOM 6380, 6991. Tiny hbr well sheltered from W in lee of Cabo Ogoño, but exposed to N/E. At ent turn hard port into S part of outer hbr to ⚓ (buoy the ⚓) or pick up buoy in about 3m. AB on S mole is feasible, but a 1m wide underwater ledge requires holding-off line to kedge ⚓. The N part of outer hbr and all the inner hbr dry.
Lts: N Mole, Fl G 3s 8m 4M; S Mole, F WR 7m 8/5M, W000°-315°, R 315°-000° showing over inshore dangers to the SE. Hr Mr ☎ (94) 6555119. **Facilities**: basic V, R, Bar.

BERMEO 43°25'·38N 02°42'·55W. Tides: see IG 11. AC 1171, 1102; SC 917, 942, 128; SHOM 6380, 6991. A busy fishing & commercial port, well sheltered from the N by a huge N mole, but swell enters in E'lies. Hbr is on NW side of the wide drying estuary of Rio Mundaca. Isla de Izaro lies off the estuary mouth, 1M ENE of hbr. Cabo Machichaco, Fl 7s 120m 24M, Siren (2) 60s, is conspic 2·75M to NW. Appr on 220° in the W sector of Rosape Dir lt, Fl (2) WR 10s 36m 7M, R108°-204°, W204°-232°, which clears Isla de Izaro and the N mole hd, Fl G 4·5s 16m 4M.
Other hbr lts: S mole hd, Fl R 3s 6m 3M; inner spur of N mole, FG 5m 2M; ent to inner basin, FG and FR. AB on N mole or on S quay of inner basin (Puerto Mayor) if FVs at sea. Or ⚓ in outer basin in 4-6m. The old inner hbr, N of Puerto Mayor, partly dries to rks and is untenable. VHF Ch 09,16. Hr Mr ☎ (94) 6186445 or mobile ☎ 908 873330. **Facilities**: AB, FW, D, Slip, C (12 ton), ME, Sh, Ice. **Town**: V, R, P, Gaz, El, Ⓔ, Ⓑ, ✉; ⇌ tourist line to Mundaca & Guernica; Ferry & ✈ at Bilbao.

BILBAO (Bilbo)   IG 12

Vizcaya 43°22'·80N 03°04'·80W (outer entrance)

CHARTS
AC 74, 1170, 1102; SC 3941, 941/2, 128; SHOM 6774, 6991

TIDES
Standard Port PTE DE GRAVE (←—); ML 2·4; Zone –0100

Times				Height (metres)			
High Water		Low Water		MHWS	MHWN	MLWN	MLWS
0000	0600	0500	1200	5·4	4·4	2·1	1·0
1200	1800	1700	2400				
Differences ABRA DE BILBAO							
–0125	–0045	–0035	–0055	–1·2	–1·2	–0·5	–0·4
PORTUGALETE (INNER HARBOUR)							
–0100	–0020	–0010	–0030	–1·2	–1·2	–0·5	–0·4

SHELTER
Excellent inside Las Arenas yacht hbr and in Getxo marina on S side of Contramuelle de Algorta opened in summer '97; or ⚓ off in 4m. Bilbao is a major industrial city but the SE part of the inner hbr is clean and adjacent to a pleasant suburb.

NAVIGATION
WPT 43°23'·10N 03°05'·34W, 310°/130° from/to outer hbr ent, 5ca. The outer ent to El Abra (inlet/roads) is formed by a W bkwtr extending 1.25M NE from Punta Lucero, a high unlit headland on the W side. The partly-finished E bkwtr extends WNW for almost 2M from Punta Galea, Fl (3) 8s 82m 19M, on the E side of El Abra. It is marked by 3 SPM lt buoys, but can be crossed with caution; approx 7m depth was found close off Pta Galea.
From the outer ent steer 127° across El Abra for 3M to round the Dique de Santurce, Fl (2) G 12s. In 1998/9 much construction work in progress on the SW side of El Abra. The head of a new bkwtr extending E from the SW shore was recorded by GPS at 43°21'·52N 03°02'·66W, not far from the fairway. After passing Contramuelle de Algorta head, Fl (4) R 14s, turn ESE for the new marina; or SE for Las Arenas yacht hbr across the inner hbr.

LIGHTS AND MARKS
W and E outer bkwtr heads are lit by Fl G 4s and Fl R 6s respectively. Other lts as chartlet. The Port Authority bldg (see chartlet) is conspic by day. Glare from industrial plants may be seen from afar at night. 2 tall power stn chimneys with R/W bands are conspic close W from root of Dique de Santurce.

RADIO TELEPHONE
Port VHF Ch 05, **12**, 16. Marina Ch 09.

TELEPHONE (Dial code 94)
Hr Mr 4637600, 4638061, ⌗ 4234700; Met via YC; Police 4246445; ⊞ 4903100; Brit Consul 4157600, 4167632.

FACILITIES
Marina Getxo (260, inc Ⓥs, 821 planned; max LOA 15m), ☎ 4911354, 4607435; 2200Ptas; opened summer 1997. Approx 3m. FW, AC, D & P, BH (50 ton), C (5 ton), ME, El.
Las Arenas Yacht hbr, (125 + 15 Ⓥ) AB, M 3m, FW, P, D, Slip, ME, BH (25 ton), C (3 ton).
Real Club Maritimo del Abra ☎ 4637600, 4638061, Bar, R; **Real Sporting Club** ☎ 4162411, Bar, R.
Services: Sh, CH, SM, El, Ⓔ.
City: (10km SE of inner hbr; use the Metro), all facilities; ≥, ✈ N of city (8km from hbr); ferry to Portsmouth (UK).

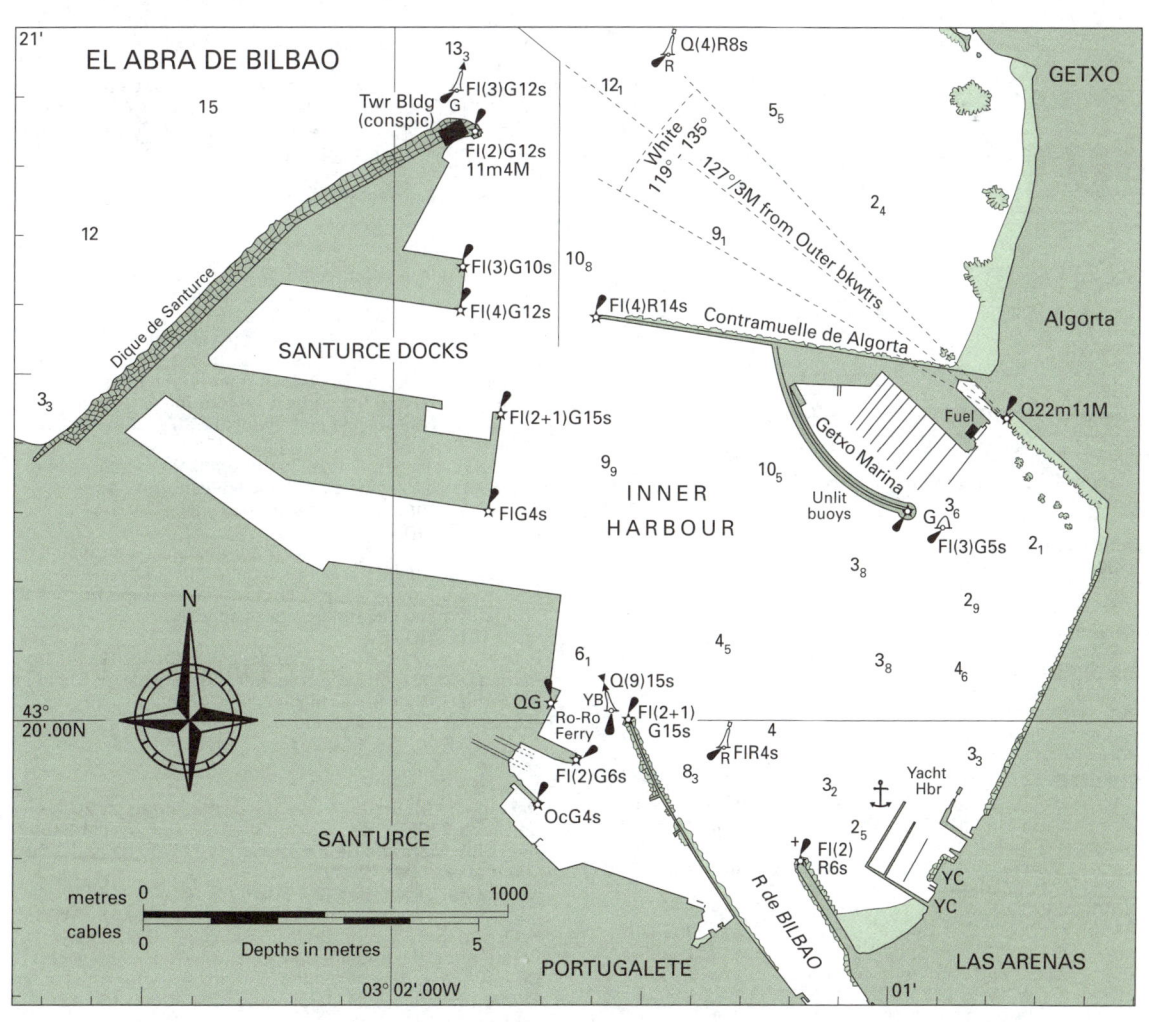

CASTRO URDIALES IG 13
Cantabria 43°22'·87N 03°12'·50W

CHARTS
AC 1171, 1170, 1102; SC 165A, 941, 128; SHOM 3542, 6991
TIDES
Standard Port PTE DE GRAVE (←); ML No data; Zone –0100

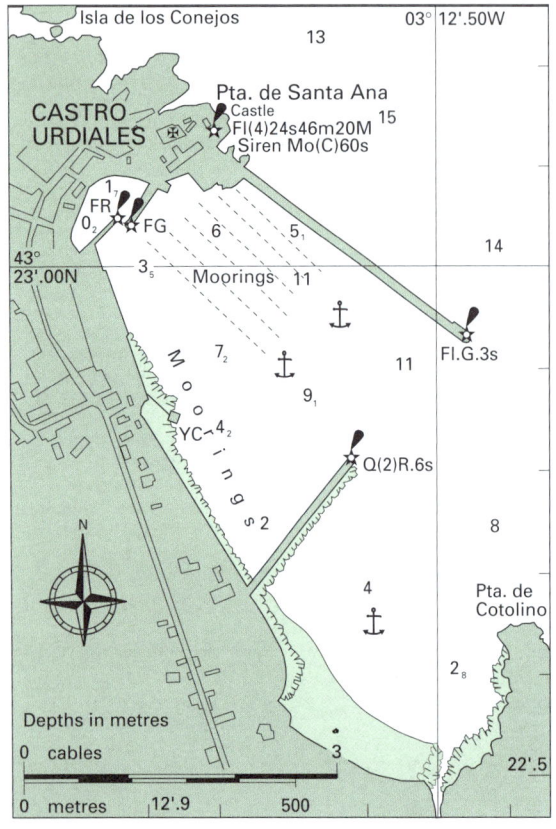

Times				Height (metres)			
High Water		Low Water		MHWS	MHWN	MLWN	MLWS
0000	0600	0500	1200	5·4	4·4	2·1	1·0
1200	1800	1700	2400				
Differences CASTRO URDIALES							
–0040	–0120	–0020	–0110	–1·4	–1·5	–0·6	–0·6
RIA DE SANTOÑA							
–0005	–0045	+0015	–0035	–1·4	–1·4	–0·6	–0·6

SHELTER
Good, except in N/E'lies when swell enters. A few ⚓s or ⚓ in 9m to seaward of the 6 lines of mooring trots which fill the N part of hbr. Many private moorings off the YC; FVs fill inner hbr. AB on N mole possible if calm and no swell.
NAVIGATION
WPT 43°23'·68N 03°11'·55W, 044°/224° from/to N Mole head lt, 1M; on 224° the mole head lts are in transit. The approach is straightforward with no hazards.
LIGHTS AND MARKS
From the NW, town & hbr are obsc'd until rounding Pta del Rabanal with its conspic cemetery. The ✠, lt ho and castle at Santa Ana are easily identified. Lts as chartlet.
RADIO TELEPHONE
Use free YC launch *Blancona* Ch 09, not your own tender.
TELEPHONE (Dial code 942)
Hr Mr 861147; Met via Hr Mr; ⌘ 861146; CG/LB 900 202 202; Dr 861640(Red Cross); Police 092.
FACILITIES
Services: D, FW, Slip, BY, AC, P (cans), ME; **Club Náutico** ☎ 861585 R, Bar. **Town** V, R, Bar, ⒷⒷ, ✉; ✈ Bilbao.

ADJACENT HARBOUR/ANCHORAGE

RÍA DE SANTOÑA Cantabria. 43°25'·90N 03°24'·67W. AC 1171, 1102; SC 24B, 940/1; SHOM 3542. Tides as above. The Lat/Long given is WPT off river mouth and on ldg lts 283·5°. The sandy estuary is dominated by Monte Ganzo (374m) on N side and is exposed to E and S. Laredo, small FV hbr, is on S side. The river narrows off Pta del Pasaje: Santoña is to the NW with 2 small, dirty basins both lit; to the S, yacht ⚓, pontoons, moorings and CN are on the W tip of Pta del Pasaje. A shallow chan runs 2M S to Colindres. Lts: Pta Pescador, Fl (3+1) 15s, is on N side of Monte Ganzo. Ldg lts 283·5°: front, Fl 2s 6m 8M; rear, Oc (2) 5s 13m 11M. Laredo, N mole FR 9m 2M. CN de Laredo ☎ (942) 605812/16, M, Slip, C (6 ton), R, Bar.

SANTANDER IG 14
Cantabria 43°27'·75N 03°47'·42W (Darsena de Molnedo)

CHARTS
AC 1155, 1102/5; SC 4011/2, 940, 127; SHOM 7365, 6991
TIDES
Standard Port PTE DE GRAVE (←); ML 2·3; Zone –0100

Times				Height (metres)			
High Water		Low Water		MHWS	MHWN	MLWN	MLWS
0000	0600	0500	1200	5·4	4·4	2·1	1·0
1200	1800	1700	2400				
Differences SANTANDER							
–0020	–0100	0000	–0050	–1·3	–1·4	–0·6	–0·6
RÍA DE SUANCES							
0000	+0030	+0020	–0020	–1·5	–1·5	–0·6	–0·6
SAN VICENTE DE LA BARQUERA							
–0020	–0100	0000	–0050	–1·5	–1·5	–0·6	–0·6
RÍA DE TINA MAYOR							
–0020	–0100	0000	–0050	–1·4	–1·5	–0·6	–0·6
RIBADESELLA							
+0005	–0020	+0020	–0020	–1·4	–1·3	–0·6	–0·4

SHELTER
Access in all weather/tides. Good shelter in both marinas: Darsena de Molnedo is central but has little space for Ⓥs; pick up ⚓ or ⚓ in 3-4m off YC, exposed to W'lies and ferry wash. Marina del Cantabrico, about 2·5M further up-river, is large, but 8km from city/facilities by taxi.
NAVIGATION
From the E, WPT 43°29'·00N 03°43'·80W, bears 055°/235° from/to Nos 2/3 chan buoys, 2M; appr between Isla de Santa Maria and Bajo Santoñuca (16·7m; breaks).
From the W, WPT 43°29'·55N 03°46'·50W (4ca E of Cabo Mayor). Thence in fair weather head SE between Isla de Mouro and Peninsula de la Magdalena (min depth 7·3m); in heavy weather keep NE of Isla de Mouro.
Punta Rabiosa ldg lts lead 236° into the well-buoyed/lit DW chan. Second ldg lts 259·5°. Lt buoys Nos 4, 5, 6 and 7 omitted for clarity. No 14 bcn tr, Fl (4) R 11s, is easily identified. Beyond the conspic oil terminal/jetty pass a SHM buoy, Fl G 4s, and an inflatable ECM buoy; after a small preferred chan buoy, Fl (2+1) G 10s, turn WSW for marina ent. The ldg marks/lts (hard to see) to marina lead 235·5° across a drying sandbank. Note: marina advises against a night entry unless already visited by day.
LIGHTS AND MARKS
The former Royal Palace, now a university, is conspic on top of Peninsula de la Magdalena. A glazed pyramid-shaped bldg about 400m W of Marina del Cantabrico is a good daymark. Lts and buoys as chartlet.
RADIO TELEPHONE
Port Authority *Santander Prácticos* VHF Ch 06, **12**, 14, 16; Marinas Ch 09 (06). Local weather is broadcast on Ch 11 at 0045 and every 4 hrs until 2045, plus 0715 and 1115.
TELEPHONE (Dial code 942)
Hr Mr 223900, ☏ 362413; ⌘ and Met via Hr Mr; Brit Consul 220000.
FACILITIES
Darsena de Molnedo (1·7m; 108 berths; apply at YC for berth), ☎ 223900, ☏ 362413, Slip, AC, D, FW, C (1·5 ton);
Real Club Marítimo de Santander ☎ 273750, ⚓ and ⚓ 50-100m off YC, Bar, R;
Marina del Cantabrico (2·2-3·1m; AB 700+, max LOA 20m; still under construction; 1226Ptas, ☎/☏ 369281, FW, C, P, D, Slip, BH (27 ton), CH, ME, Sh, El, Ⓔ, Gaz, R, Bar. Plenty of berths and adequate security to leave a yacht.
City: All amenities; ⇌ & bus; Brittany ferry terminal to Plymouth is 6ca W of Darsena de Molnedo; ✈ (Parayas).

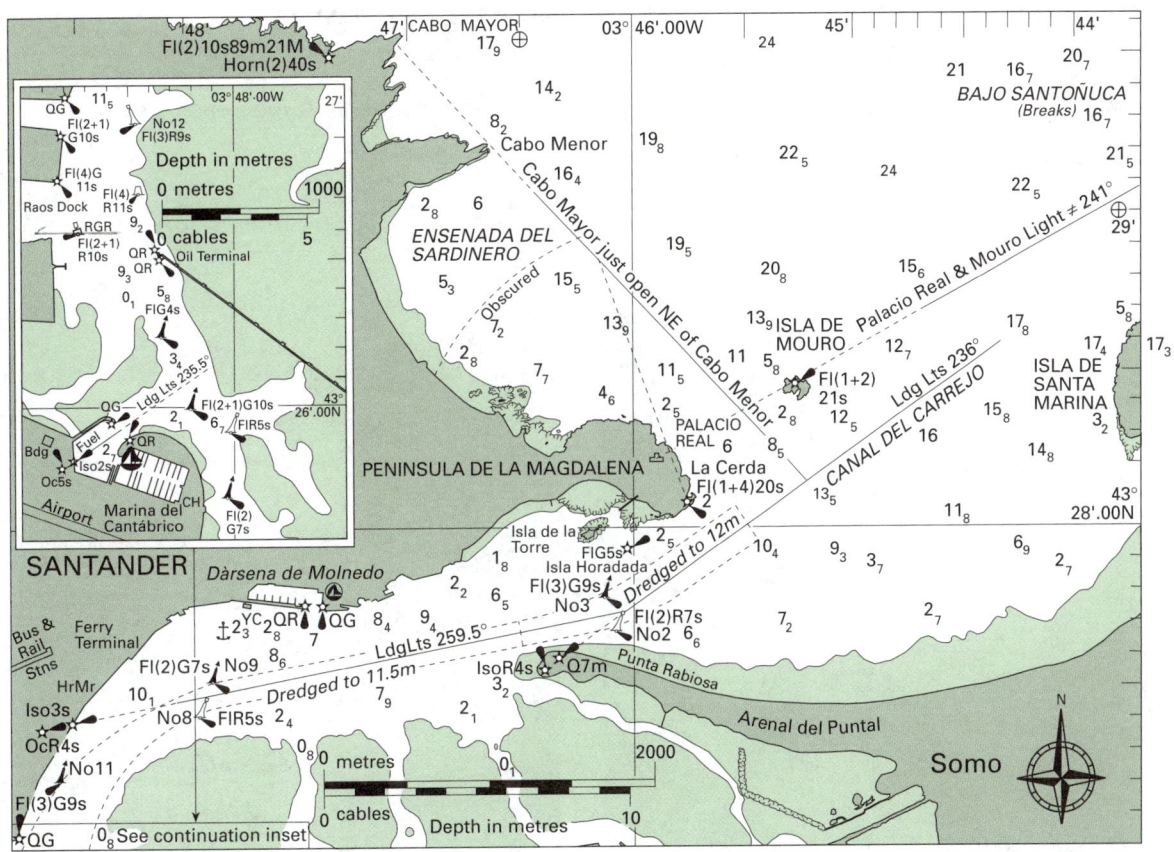

MINOR HARBOURS BETWEEN SANTANDER AND GIJON

SAN VICENTE DE LA BARQUERA, Cantabria. 43°23'·72N 04°23'·07W. AC 1150, 1105; SHOM 6381, 6991; SC 4021, 938. Tides see IG 14. Fishing hbr with good shelter in large shallow estuary, access HW±3; but HW±1 is advised if there is any swell at the entrance. Beware shallow bar and tidal streams 3-4kn. Appr on 225° to enter between Isla Peña Menor to stbd and hd of training wall to port. Beyond hotel and ☆ FG on post among pine trees keep to the NW bank; drying areas to the SE have encroached on the chan. AB on fish quay is possible or ⚓ near ship mooring buoy close to root of trng wall. Moorings by the rebuilt bridge and castle are awkward and subject to traffic noise H24. Lights: Punta de Silla, Oc 3·5s 42m 13M; Horn Mo (V) 30s. Isla Peña Menor, Fl WG 2s 7m 7/6M, G175°-235°, W235°-045°. Hd of trng wall, Fl (2) R 8s 8m 5M. Hr Mr ☎ (942) 710009; CG 900 202 202; Met 232405; Dr 712450; Police 710288. **Facilities:** Fish quay FW, ME, C, SM, D pump is for FVs only. **Town:** R, V, Bar, P & D (cans), Ⓑ, ✉; ✈ (Santander 62km).

RIBADESELLA, Asturias. 43°28'·16N 05°04'·00W. AC 1150, 1105; SC 4031, 937, 127; SHOM 6381, 6691. Tides, see IG 14. Good shelter and easy appr from about HW-1 to HW, but do not attempt in strong onshore winds when seas and swell break on the bar. Depths on the bar reported as 2m, but less after NW gales. Bajo Serropio, 1100m NNE of ent, breaks in heavy seas. To E of ent smooth, dark, steep cliffs and a chapel are distinctive. To the W a wide beach is easily identified. Keep 25m S of promenade all the way up to town; dries to stbd. AB on the NE/E quays in 2-3m before a low bridge. Lts: Somos lt ho (8ca W), Fl (1+2) 12s 113m 21M. Quay hd (Pta del Caballo), Fl (2) R 6s 10m 5M, vis 278·4°-212·9°. Hr Mr ☎ (98) 860207. **Facilities:** FW, Slip, D, P (cans), ME, CH, YC. **Town:** V, R, Bar, ⇌.

MINOR HARBOURS BETWEEN GIJON AND RIBADEO
(See overleaf)

AVILÉS, Asturias. 43°35'·80N 05°57'·00W. AC 1133, 1151; SC 4052, 935; SHOM 7361, 5009. Tides, see IG 15. A port of refuge, except in >F9 W/NW'lies, but also an industrial, commercial and fishing hbr. The ent, S of Pta del Castillo lt (Oc WR 5s 38m 20/17M) and N of conspic white beach, is open to W/NW but well sheltered from N'lies. The ent chan runs ExS for 7ca then turns S for 2M. Best ⚓ is in 5-6m on E side of chan, just S of the bend. The chan is well marked/lit all the way up-hbr. Port VHF Ch 06, 09, 12, **14**, 16. Hr Mr ☎ (985) 563013. No yacht facilities per se, but usual FV supplies and domestics. Club Náutico at Salinas.

CUDILLERO, Asturias. 43°33'·90N 06°06'·80W. AC 1108; SC 934/5, 126a; SHOM 5009. Tides, use San Estaban de Pravia; see IG 15. Good shelter in large modern basin to the NW of entrance and village, but any N'ly swell can make entry and exit difficult. From the W keep 5ca offshore until clear of Islotes las Colinas, small islands NW of ent. Appr with Punta Rebollera lt ho, Oc (4) 16s 42m 16M, brg 200°. The rocky ent, marked by FR and FG lts, opens up only when close to. Turn 90° stbd through narrow inlet into basin. AB on N'ly of 3 pontoons on N wall, or ⚓ off these pontoons in 8-9m. FVs occupy the NW end of the basin. **Facilities:** FW, D from FV quay, P (cans), ME, Slip, small commercial shipyard. **Village:** 5 mins by dinghy, 25 mins on foot; V, R, Bar, ✉.

LUARCA, Asturias. 43°33'·00N 06°32'·10W. AC 1133, 1108; SC 933/4, 126a; SHOM 6381, 5009. Tides, see IG 15. Good shelter in the outer hbr protected from all but N'lies; moor fore and aft to the E bkwtr or alongside if no swell. No room to ⚓ except outside the E bkwtr in settled weather. Narrow chan leads SE to inner hbr dredged 2m (prone to silting). Pontoons in SE corner are full of local craft; temp'y berth on NW side may be possible. Appr on the ldg lts/marks 170° to clear rky shoals either side. The E bkwtr lt ho is conspic and almost in transit with the ldg marks. Lts: Cabo Busto, Fl (4) 20s 84m 25M, is 3M ENE of hbr ent. Pta Blanca, Oc (3) 15s 63m 14M, siren Mo (L) 30s, W □ tr and house on prominent headland 300m ENE of hbr ent; also conspic ✠. Ldg lts 170°: front, Fl 5s 18m 2M; rear, Oc 4s 25m 2M; both on thin W pylons, R bands. W bkwtr hd, Fl (3) G 9s 7m 5M. E bkwtr hd, Fl (3) R 9s 22m 5M, ○ concrete tr. Inner spur, Fl (4) G 11s 6m 1M. RCN ☎ (98) 5640130, Bar. **Facilities:** FW, AC, D from FV quay, P (cans), ME, CH, Slip, C (8 ton). **Town:** V, R, Bar, Ⓑ, ✉, ⇌.

GIJÓN IG 15
Asturias 43°32'·82N 05°40'·06W
CHARTS
AC 1151, 1105, 1108; SC 4042, 935, 127; SHOM 6381, 5009

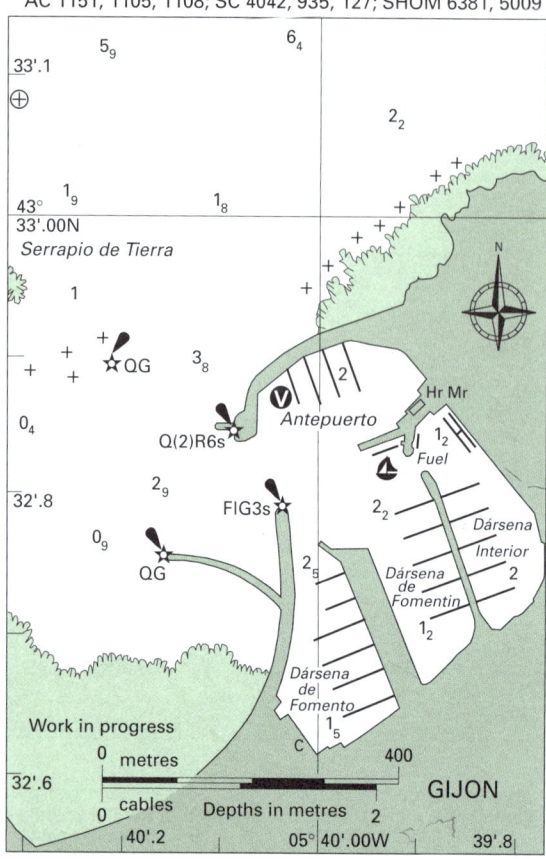

TIDES
Standard Port PTE DE GRAVE (←); ML 2·3; Zone –0100

Times				Height (metres)			
High Water		Low Water		MHWS	MHWN	MLWN	MLWS
0000	0600	0500	1200	5·4	4·4	2·1	1·0
1200	1800	1700	2400				
Differences GIJON							
–0005	–0030	+0010	–0030	–1·4	–1·3	–0·6	–0·4
LUANCO							
–0010	–0035	+0005	–0035	–1·4	–1·3	–0·6	–0·4
AVILÉS							
–0100	–0040	–0015	–0050	–1·5	–1·4	–0·7	–0·5
SAN ESTABAN DE PRAVIA							
–0005	–0030	+0010	–0030	–1·4	–1·3	–0·6	–0·4

SHELTER
Excellent in marina, access H24, depths 1·2 - 2·8m in four basins. Marina (Puerto Local or Muelles Locales) is 1M SE of Puerto del Musel, the huge conspic industrial port.
NAVIGATION
WPT 43°33'·08N 05°40'·30W, 330°/150° from/to marina ent, 3ca. Banks in the appr's are marked by cardinal lt buoys. From E or W, start about 1ca E of lt ho, Fl G 3s 22m, on Puerto del Musel's outer bkwtr hd; thence steer 180°/1·2M to the WPT. From WPT, steer about 150° towards marina ent, passing close E of Sacramento G tr, QG, (marks drying reef). Initially N mole obscures S mole lt, Fl G 3s.
LIGHTS AND MARKS
Cabo de Torres, Fl (2) 10s 80m 18M, and conspic W tanks are 2M NW of marina. Many nav/shore lts in P. del Musel.
RADIO TELEPHONE
Marina VHF Ch 09. Puerto del Musel Ch 11, 12, **14**, 16.
TELEPHONE (Dial code 98)
Hr Mr 535 41 28; ⌖ & Met via marina.
FACILITIES
Marina (750+120 Ⓥ), ☎ 534 45 43, 🛟 535 00 27, AC, FW, P, D, Slip, C (10 ton),YC; **Services:** ME, El, Sh, CH, Ⓔ, ⌂.
Town: V, R, Bar, ✉, Ⓑ, ⇌; ✈ Oviedo (26 km).

RÍA DE RIBADEO IG 16
Lugo 43°32'·53N 07°02'·10W
CHARTS
AC 1122, 1108; SC 550A, 932, 126; SHOM 6383, 5009

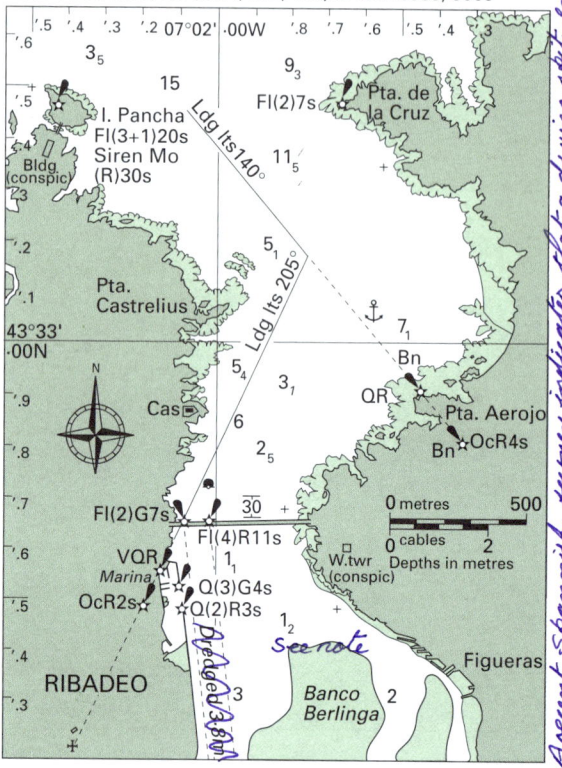

A recent Spanish survey indicates that a drying spit extends from the NW tip of Banco Berlinga to within 140m of the bridge. At the E end of the bridge a drying shoal extends SW and almost along the above spit. The Chan to Figueras is now impassable and shoaler than before. Access to the marina is not affected by the above.

TIDES
Standard Port PTE DE GRAVE (←); ML No data; Zone –0100

Times				Height (metres)			
High Water		Low Water		MHWS	MHWN	MLWN	MLWS
0000	0600	0500	1200	5·4	4·4	2·1	1·0
1200	1800	1700	2400				
Differences LUARCA							
+0010	–0015	+0025	–0015	–1·2	–1·1	–0·5	–0·3
RIBADEO							
+0010	–0015	+0025	–0015	–1·4	–1·3	–0·6	–0·4
RÍA DE VIVERO							
+0010	–0015	+0025	–0015	–1·4	–1·3	–0·6	–0·4
SANTA MARTA DE ORTIGUEIRA							
–0020	0000	+0020	–0010	–1·3	–1·2	–0·6	–0·4

SHELTER
Good in marina (approx 2m), close S of W end of bridge (30m clnce). Visitors berth alongside outer mole. Nearby ⚓s: close N of outer ldg lts; off Figueras (appr near HW due to shallow chan); and off Castropol in 4m.
NAVIGATION
From WPT 43°34'·00N 07°01'·93W (off chartlet), appr by day with W tr (conspic at E end of bridge) brg 170°, clearing Pta de la Cruz, to intersection of outer and inner ldg lines. Thence follow inner ldg line 205° past Pta Castrelius and through W span of bridge. Caution: extensive sandbanks obstruct the E side of the ría, both N and S of the bridge. Some depths may be less than charted on AC 1122. Tides run hard through bridge.
LIGHTS AND MARKS
All lts in the ría as shown. Daymarks on both the 140° and 205° ldg lines are W trs with R ◊. Conspic R onion-shaped ✠ twr in the town is almost aligned on the inner ldg line.
RADIO TELEPHONE
VHF Ch 16 H24.
TELEPHONE (Dial code 982)
Hr Mr 131144; ⌖ & Met via Hr Mr.
FACILITIES
Marina (30+ few visitors), ☎/🛟 131144; fee is 12ptas x m² (ie LOA x beam); FW, Slip, P & D (cans), ME, C (8 ton), Sh; **Club Náutico del Eo** R, Bar. **Town:** Gas, V, R, Bar, Ⓑ, ✉, ⌂.

HARBOURS SSE OF PUNTA DE LA ESTACA DE BARES

RÍA DE VIVERO, Lugo. 43°41'·00N 07°36'·12W. AC 1122, 1108; SC 4082, 931, 126a; SHOM 6383, 5009. Tides: IG 17. Easy ent, 7ca wide, between Pta Socastro, Fl G 7s 18m 5M, and Pta de Faro, Fl (2) R14s 18m 5M. Good ⚓ in E'lies off Isla Insua d'Area, 1·3M NE of Cillero. Steer 195°/2M for Cillero bkwtr hd, FR 9m 4M; avoid the N and E quays of this major FV hbr. AB on rough quay to the SSW or ⚓ off Playa de Covas in 6-7m. 6ca up-river a marina basin (3m) lacks pontoons; opening date *mañana*. VHF Ch 15, 16. Hr Mr ☎ (982) 560074, 📞 560410; YC ☎ 561014; Dr 561202; Police 562922; **Facilities**: D & P, FW, ME. **Town** (Vivero): V, R, Bar, Ⓑ, ✉, ⇌, Ⓗ (Burela 20km, ☎ 982-589900).

RÍA DEL BARQUERO. 43°45'·00N 07°40'·5OW. AC 1122; SC 18A; SHOM 6383. Tides: IG 17. No shelter from NE. Small hbrs at Bares, El Barquero (☎ 981-414002) and Vicedo. Lts: Pta de la Barra, Fl WRG 3s; Pta del Castro, Q (2) R 6s.

RÍA DE CEDEIRA   IG 17

La Coruña 43°39'·50N 08°03'·80W

CHARTS
AC 1122, 1111; SC 930, 41A; SHOM 6383, 5009, 3007
TIDES
ML No data; Zone –0100. Interpolate between SANTA MARTA DE ORTIGUEIRA (IG 16) and EL FERROL (IG 18)
SHELTER
Very good; a pleasant refuge for yachts trying to weather Cabo Ortegal in a NE'ly. ⚓ about 500m E of the modern FV port in 3-4m, clear of the fairway. The E/SE end of the bay shoals rapidly toward the drying river mouth.
NAVIGATION
WPT 43°41'·25N 08°05'·60W (7½ca off chartlet), 335°/155° from/to Pta Promontorio lt ho, 2·35M. From N keep about 1M off Pta Candelaria until the ría ent opens. From the S a similar offing will clear rks off Pta Chirlateira. Appr on 155° transit of Pta Promontorio lt ho and Pta del Sarridal; when abeam Pta Chirlateira alter 161° to pass midway between Pta del Sarridal and Piedras de Media Mar, W lt tr conspic in centre of ría. Then steer E past bkwtr head.
LIGHTS AND MARKS
Pta Candelaria, Fl (3+1) 24s, is 2·6M NE of the WPT. Pta de la Frouxeira, Fl (5) 15s, is 6M to the SW. Piedras de Media Mar, Fl (2) 5s 12m 4M, commands the ría and can be passed either side. *Pta del Sarridal Oc WR 6s 39m 11M*
RADIO TELEPHONE
No VHF. MF *Cedeira Cofradia* 1800kHz, 2182.
TELEPHONE (Dial code 981)
Hr Mr 480389; ⌗ & Met via Hr Mr.
FACILITIES
Hbr: FW, Slip, ME, El, Sh, L just N of ⚓ symbol. **Town**: Ⓔ, D & P (cans), V, R, Bar, Ⓑ, ✉, ⓘ, ⇌ & ✈ La Coruña.

EL FERROL   IG 18

La Coruña 43°28'·62N 08°14'·50W (Curuxeiras)

CHARTS
AC 1115, 1111; SC 4122/3/4, 412A, 929; SHOM 6665, 3007
TIDES
Standard Port PTE DE GRAVE (⟵); ML 2·2; Zone –0100

Times				Height (metres)			
High Water		Low Water		MHWS	MHWN	MLWN	MLWS
0000	0600	0500	1200	5·4	4·4	2·1	1·0
1200	1800	1700	2400				
Differences EL FERROL							
–0045	–0100	–0010	–0105	–1·6	–1·4	–0·7	–0·4

SHELTER
Good inside the narrow, but straightforward ent to ría. El Ferrol is a naval and commercial port with limited yacht facilities. Note: chartlet shows only the central part of the hbr, ie between the narrows to the W and the naval and commercial docks to the E. Caution: New Spanish charts show depths up to 2m less than charted in some areas. A small marina at La Graña may have AB or ⚓/moor off. Dársena de Curuxeiras is central: AB in 2m on NE side on stone walls and prone to ferry wash; 3 pontoons at the head of this inlet are not available to visitors. ⚓s, from seaward: off Cariño and Pereiro on N shore of ent chan and at Baño and Mugardos inside the ría (see chartlet).
NAVIGATION
WPT 48°27'·15N 08°20'·00W, 238°/058° from/to S. Cristobal Dir lt, Oc (2) WR 10s, 1·6M. In bad weather avoid the outer banks (Tarracidos, Cabaleiro and Leixiñas) to the N and W.
LIGHTS AND MARKS
C. Priorino Chico, Fl 5s 34m 23M, marks N side of ría ent. Inner ldg lts 085° to the narrows; front Fl 1·5s, rear Oc 4s; both W trs. Chan is deep and adequately buoyed/lit.
RADIO TELEPHONE
Port Control *Ferrol Prácticos* VHF Ch 14 (H24).
TELEPHONE (Dial Code 981)
Hr Mr 352103, 📞 353174; ⌗ & Met via Hr Mr.
FACILITIES
Marina Terramar at La Graña (approx 50), FW, Slip, AC, ME, P & D (cans), El, C (1·5 ton), Sh; bus every ½hr to the city (20 mins); **Club Náutico. Curuxeiras** ☎ 321594; Slip, FW, D, P (cans), ME, El, CH, C (8 ton). **City**: All needs, ⇌; ✈ (La Coruña).

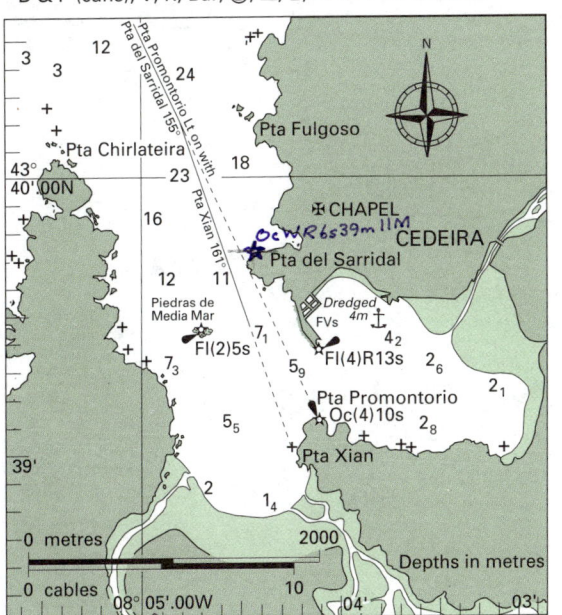

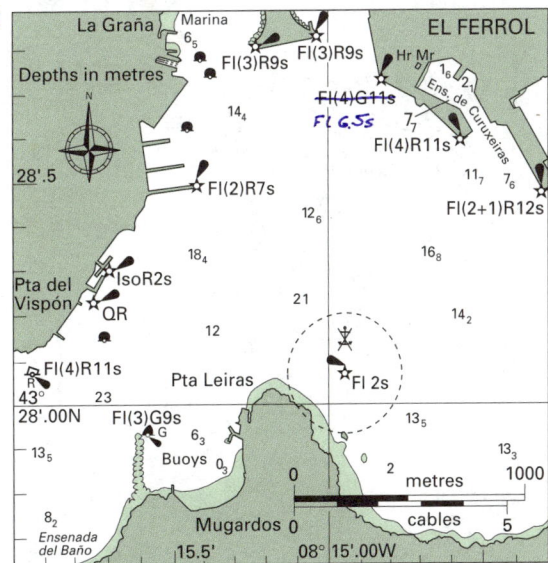

ADJACENT HBRS IN RÍAS DE ARES AND DE BETANZOS

ARES 43°25'·50N 08°14'·00W. AC 1114, 1111; SC 4125, 412A. Wide, sheltered bay with modern FV basin in SW corner, bkwtr hd Fl (3) R 9s. Possible AB in basin or ⚓ off in 3m or to E of Pta Modias, clear of mussel beds.

FONTAN 43°21'·70N 08°14'·35W. Charts as above. Good shelter; FVs berth inside N mole. 3 yacht pontoons at W side, but max LOA 9m. ⚓ in 3m inside bkwtrs, clear of fairway and La Pulgueira rk, Fl (4) R 11s. Good CH.

LA CORUÑA   IG 19

La Coruña 43°22'·13N 08°23'·10W (marina)

CHARTS
AC 1114, 1111; SC 4126, 412A, 929; SHOM 6665, 3007

TIDES
Standard Port PTE DE GRAVE (←); ML 2·2; Zone –0100

Times				Height (metres)			
High Water		Low Water		MHWS	MHWN	MLWN	MLWS
0000	0600	0500	1200	5·4	4·4	2·1	1·0
1200	1800	1700	2400				
Differences LA CORUÑA							
–0110	–0050	–0030	–0100	–1·6	–1·6	–0·6	–0·5

SHELTER
Good at marinas, protected from wash by unlit floating wavebreak close NE of Cas S. Anton. Two N'ly pontoons are Sporting Club (Casino); S'ly 2 are Real Club Náutico. Speed limit 3kn. Both YCs offer ⚓s; or ⚓ further out on possible foul ground. At Dársena de la Marina, 2 small pontoons close W of old RCN bldg may be negotiable.

NAVIGATION
WPT 43°23'·28N 08°22'·03W, 011°/191° from/to head of Dique de Abrigo, Fl G 5s, 1·35M. Note the 108° & 182° ldg lines meet at this WPT. Avoid Banco Yacentes, about 1M NW of WPT, between the ldg lines (off chartlet).

LIGHTS AND MARKS
Conspic marks: Torre de Hércules, Fl (4) 20s; 5·3M SW of which is power stn chy (218m); twin white twrs (85m, R lts, hbr control) at root of Dique de Abrigo, next to marina. Pta Fiaiteira Ldg Its 182°: front, Iso WRG 2s; rear, Oc R 4s.

RADIO TELEPHONE
Port Control *Dársena Radio Torre Hércules* VHF Ch 12; VTS 13; Marinas 09; LB 73.

TELEPHONE (Dial code 981)
Sporting Club (Casino) ☎ 209007, 📠 213953; Real Club Náutico ☎ 207910, 📠 203008; Hr Mr 226001, 📠 205862; # & Met via YCs; ⊞ 277905; Dr 287477.

FACILITIES
Marinas: (250+ some Ⓥ), 2500ptas, AB (F&A), FW, AC, P, D, M, Slip, BH (30 ton), ME, El, C (1 ton), Sh, ▢, CH, SM, Ⓔ. **City:** all amenities; 🚆, ✈ (international via Santiago).

RÍA DE CAMARIÑAS   IG 20

La Coruña 43°07'·62N 09°10'·88W (Camariñas)

CHARTS
AC 1113, 1111, 3633; SC 9272, 927, 928; SHOM 7559, 3007

TIDES
Standard Port PTE DE GRAVE (←); ML No data; Zone –0100

Times				Height (metres)			
High Water		Low Water		MHWS	MHWN	MLWN	MLWS
0000	0600	0500	1200	5·4	4·4	2·1	1·0
1200	1800	1700	2400				
Differences RÍA DE CORME							
–0025	–0005	+0015	–0015	–1·7	–1·6	–0·6	–0·5
RÍA DE CAMARIÑAS							
–0120	–0055	–0030	–0100	–1·6	–1·6	–0·6	–0·5

SHELTER
Good in Camariñas on 2 YC pontoons, except in E/NE'lies; or ⚓ to S, inside bkwtr. Mugia (43°06'·39N 09°12'·60W) is solely a FV hbr, sheltered from all but E/SE'lies; ⚓ outside or in lee of bkwtr. Possible temp'y AB with fender board.

NAVIGATION
WPT 43°07'·60N 09°13'·75W, 288°/108° from/to Pta de Lago Dir lt, 2·9M. From N or W beware Las Quebrantes shoal on which sea breaks. The N'ly appr is unlit.

LIGHTS AND MARKS
C. Villano, Fl (2) 15s, with 23 conspic wind turbines close SE, is 2·2M N of ent to ría. Pta de la Barca, Oc 4s 13m 7M, marks S side. From SW, appr on 080° ldg lts, as chartlet. Pta de Lago lt, Oc (2) WRG, W sector 107·8°-109·1°, leads 108·5° into the ría. Turn for Camariñas when bkwtr bears 030°. For Mugia, turn when bkwtr head bears 190°.

RADIO TELEPHONE
Camariñas YC VHF Ch 09.

TELEPHONE (Dial code 981)
CNC 737130; # & Met via CNC; Mugia Hr Mr 742030.

FACILITIES
Club Náutico Camariñas (60 + Ⓥ), ☎ 737130, 📠 736325; F&A, 1160Ptas. **Services** (both hbrs): Slip, AC, D & P (cans), FW, ME, C, Sh.
Villages: V, R, Bar, ✉, Ⓑ; Bus to La Coruña, ✈ Santiago de Compostela.

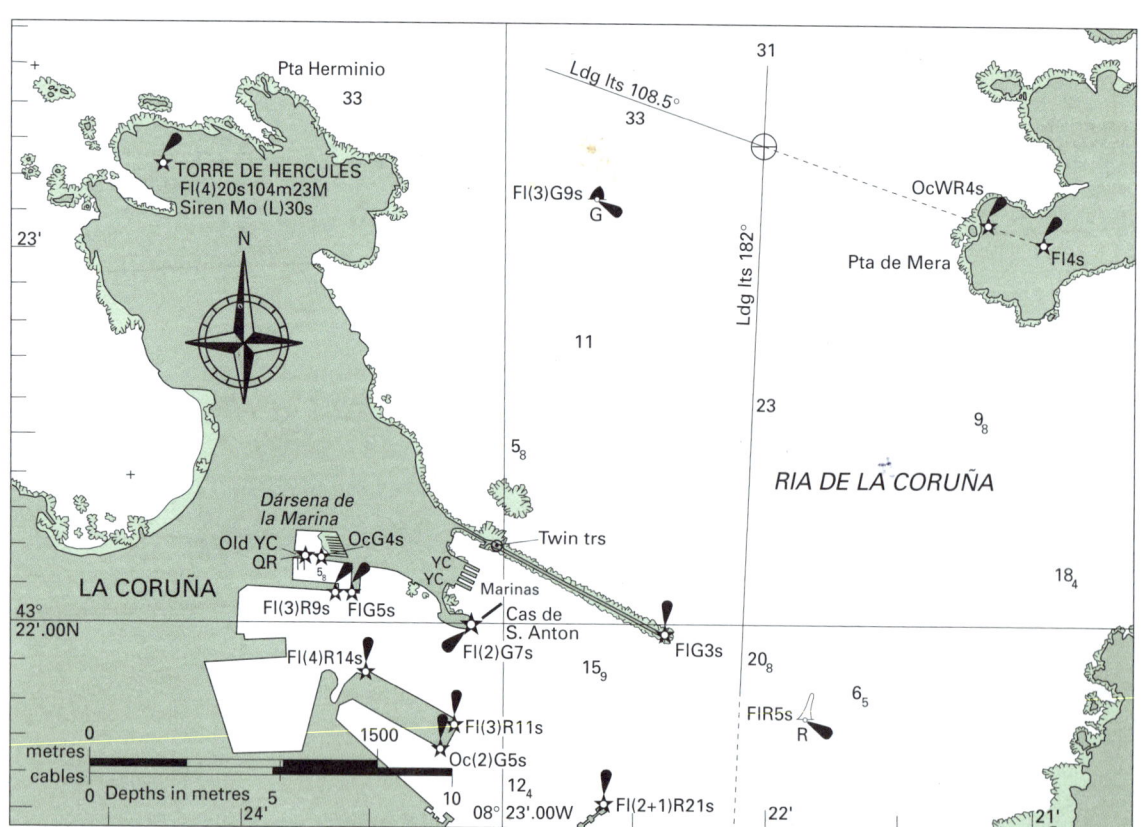

RÍA DE MUROS IG 21

La Coruña 42°43'·50N 09°04'·00W (Ent WPT)

CHARTS
AC 1756; SC 9264, 9260, 926; SHOM 5441, 3007
TIDES
Standard Port LISBOA (→); ML No data; Zone −0100

Times				Height (metres)			
High Water		Low Water		MHWS	MHWN	MLWN	MLWS
0500	1000	0300	0800	3·8	3·0	1·4	0·5
1700	2200	1500	2000				
Differences CORCUBION							
+0055	+0110	+0120	+0135	−0·5	−0·4	−0·2	0·0
MUROS							
+0050	+0105	+0115	+0130	−0·3	−0·3	−0·1	0·0

SHELTER
Good at Muros (42°46'·71N 09°03'·25W), but 2 pontoons in the inner hbr are full of locals. ⚓ off on patchy holding (2 or 3 shots may be needed) or at Ens de San Francisco. Noya, at the head of the ría, is only accessible to shoal draft. Portosín marina (3·5-8m) is at 42°45'·96N 08°56'·77W. El Son has difficult access and no yacht facilities.
NAVIGATION
WPT 42°43'·50N 09°04'·00W, 1M SSE of Pta Queixal lt and 2·3M NW of Pta Castro lt, Fl 5s (off chartlet). From the N keep to seaward of Bajo de los Meixidos and Los Bruyos which lie up to 6M WNW of Pta Queixal. The ría is mostly deep and clear, but beware mussel rafts off the N shore.
LIGHTS AND MARKS
Mte Louro is a conspic conical hill at the W side of ent. At the head of the ría Isla Quiebra is a good mark. At night Pta Queixal and Cabo Reburdino light the NW shore and Pta Castro and Pta Cabeiro the SE.
RADIO TELEPHONE
Club Náutico Portosín VHF Ch 09.
TELEPHONE (Dial code 981)
Portosín YC 766583; ⌘ & Met via YC. Porto do Son 821307.
FACILITIES
Muros: FW, D (not near LW), P (cans), V, R, Bar, Ⓑ, ✉; ⇌ and ✈ Santiago de Compostela (40 km by bus via Noya). **Club Náutico Portosín marina** (200+ some Ⓥ), ☎ 766583, 🛥 766389, F&A Ptas 2025, Slip, BH (32 ton), AC, D, P, FW, ME, El, C (3 ton), Sh, R, Bar, 🚿; **Portosín**: CH, V, R, Bar, ✉.

FINISTERRE VTS. Finisterre MRCC, c/sign *Finisterre Trafico*, provides a reporting service (voluntary for yachts) on Ch 11 16. Its area lies from the coast out to 10°10'W between brgs of 310° from C. Villano and 255° from C. Finisterre, including the TSS and ITZ. The service provides: listening watch for distress H24; radar info on request; broadcasts in Spanish & English of: traffic; nav warnings every 4hrs from 0033; and weather info every 4hrs from 0233UT.

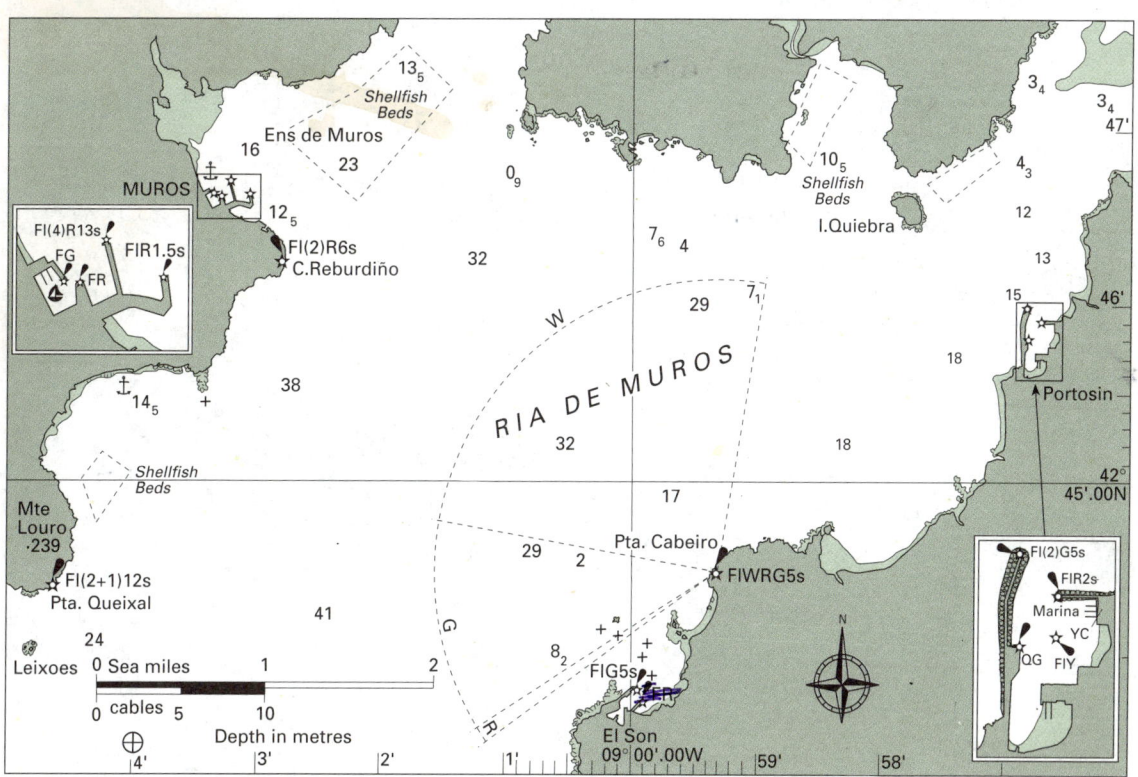

RÍA DE AROSA IG 22

La Coruña (NW)/Pontevedra (SE). ⊕ 42°26'·50N 08°59'·00W

CHARTS
AC 1768, 1757, 3633; SC 9262/3, 4153; SHOM 5536, 3007

TIDES
Standard Port LISBOA (⟶); ML 2·05; Zone −0100

Times				Height (metres)			
High Water		Low Water		MHWS	MHWN	MLWN	MLWS
0500	1000	0300	0800	3·8	3·0	1·4	0·5
1700	2200	1500	2000				
Differences VILLAGARCIA							
+0040	+0100	+0110	+0120	−0·3	−0·2	−0·1	0·0

SHELTER
Shelter can be found from most winds. Santa Eugenia and Caramiñal on the NW shore and Villagarcia, principal hbr/town, at the NE corner have marinas (see insets). At the head of the ría, Rianjo has a yacht pontoon in about 3m inside the hbr. FV hbrs (clockwise from ent) include: Aguino, Puerto Cruz, Carril, Villanueva, S. Julian (Isla Arosa), Cambados (N'ly of two hbrs) and San Martin del Grove (2 pontoons, only for very small craft).

NAVIGATION
WPT 42°26'·50N 08°59'·00W, 198°/018° from/to Is Rua (lt), 6·8M. Isla Sálvora (lt) at the mouth of the ría should be left to port; the rocky chans between it and C. Corrubedo 7M to the NNW are best not attempted by strangers. Ría de Arosa, the largest ría (approx 14M x 7M), is a mini-cruising ground with interesting pilotage and dozens of anchorages to explore. Its coast is heavily indented and labyrinthine. AC 1768 or SC 9262/3 are essential. The fairway up to Villagarcia is buoyed/lit, but to either side are many mussel rafts (*viveros*), often not marked or lit. Some of the lesser chans require careful pilotage to clear shoals/rocks. The low bridge between Isla de Arosa and the E shore prevents passage.

LIGHTS AND MARKS
Principal lts/marks are on chartlet as scale permits. Isla Rúa, a prominent rky islet with lt ho, is a key feature.

RADIO TELEPHONE
Villagarcia, Caramiñal and Santa Eugenia VHF Ch 09 16.

TELEPHONE (Dial code 986)
Villagarcia Hr Mr 500232/8, ☎ 507923; ⌗ & Met via Hr Mr; Caramiñal and Sta Eugenia marinas, see below.

FACILITIES
Villagarcia marina (416+ Ⓥ on 1st pontoon to stbd); ☎ & ⌒, as above; AC, FW, Slip, BH (35 ton), P, D, ME, El, Sh, C (70 ton), SM, Ⓔ, YC. **Town**: CH, V, R, Bar, Ⓑ, ⌧, ⌥, ⇌; ✈ Santiago de Compostela (53km).
Caramiñal marina (150 F&A on ⚓ pontoons in 3·5m); FW, AC, P & D (cans); **Club Náutico** ☎ & ⌒ (981) 830970;
Sta Eugenia marina ☎ (981) 873801, ⌒ 873290; (70 F&A on T-shaped pontoons in 4m); FW, AC. **Town**: all facilities.
Rianjo: pontoon in 3m, FW, AC; ☎ 860477. No sign of planned marina in July '97.

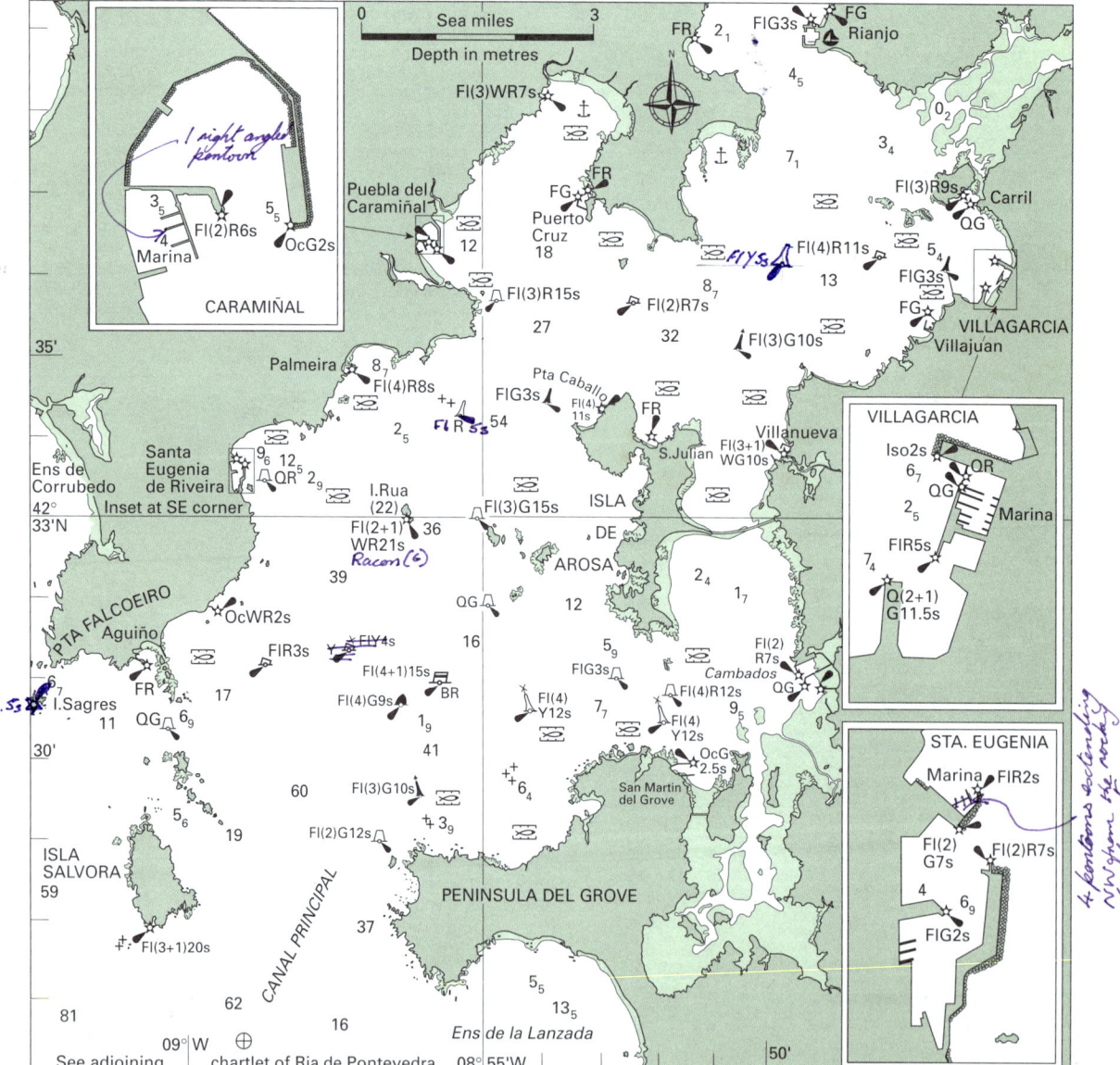

RÍA DE PONTEVEDRA　　IG 23

Pontevedra 42°22'·00N 08°50'·00W (centre of ría)

CHARTS
　　AC 1758, 3633; SC 9251, 4161/2/3, 925; SHOM 5517, 3007

TIDES
　　Standard Port LISBOA (→); ML 1·9; Zone –0100

Times				Height (metres)			
High Water		Low Water		MHWS	MHWN	MLWN	MLWS
0500	1000	0300	0800	3·8	3·0	1·4	0·5
1700	2200	1500	2000				
Differences MARIN							
+0050	+0110	+0120	+0130	–0·5	–0·4	–0·2	0·0

SHELTER
　　Good shelter can be found from any wind. Islas Ons and Onza off the mouth of the ría offer a barrier to wind and swell. Access to Pontevedra, the provincial capital, is restricted by a shallow chan and low bridge. Marin is a naval base and commercial port of little interest to yachts. Other hbrs and anchs, clockwise from NW ent, include, on the N shore: Porto Novo, ~~Sangenjo~~, Rajó and Combarro; on the S shore: ~~Aguete~~, Bueu and Aldán. Anchorage on E side of Isla Ons. See details under FACILITIES.

NAVIGATION
　　From WPT 42°18'·00N 08°56'·88W, (which also lies on the 129° ldg lts for Ria de Vigo) track 053° to enter via the main chan, Boca del Sudoeste, into centre of ria between C. de Udra and Pta Cabicastro. From N, transit Paso de la Fagilda on 130°, or Canal de Los Camoucos on 175°; both chans need care. Isla Tambo is a restricted military area; landing prohib.

LIGHTS AND MARKS
　　Lts and marks as chartlet. Isla Ons is a steep, rky island with conspic lt ho, Fl (4) 24s 126m 25M, octagonal tr.

RADIO TELEPHONE
　　Marin VHF Ch 12 16.

+ Sangenjo has new marina. Aguete marina is not yet fully completed.

TELEPHONE (Dial code 986)
　　See below for ☎ numbers at hbrs.

FACILITIES
　　Porto Novo: Full of moorings, no space to ⚓ inside the hbr. Find a vacant ⚓ or ⚓ NE of ent in about 6m. Small pontoon on the N mole is suitable for landing only. Keep clear of FVs and ferries on S mole, Fl (3) R 6s. CN ☎ 723266, FW, D, CH, ME, V, R, Bar, Ⓑ. Ferry to Isla Ons.
　　Sangenjo: ~~Small FV hbr, full of moorings.~~ Enter between mole hd, Fl (4) R 12s, and unlit W tr, G top, marking shoal. CN ☎ 720059, FW, D, C (5 ton), ME, V, R, Bar, Ⓑ, ⛽.
　　Rajó: Temp'y ⚓ in about 5m off the bkwtr hd, Fl (2) R 8s.
　　Combarro: ⚓ in about 3m to SE of bkwtr hd, Fl (2) R 8s. A new mole 1ca S of the old bkwtr gives added shelter; the lt on hd is inop. The bay is generally shallow; the village is noted for *horreos* and tourists. Possible security risk.
　　Pontevedra: Enter by dinghy near HW between training walls, Fl G/R 5s. The N side of chan avoids shoals and old bridge foundations in mid-channel between overhead cables and motorway bridge (both 12m clearance). Small private marina in Rio Lérez beyond. Town: all facilities.
　　Aguete: New marina in pleasant bay, but the 2 pontoons S of floating wavebreak were not in use July '97; many ⚓s. The YC bldg resembles superstructure of a liner. Do not round mole hd, lit, too close due to offlying rks to SW/W.
　　Marina (approx 100 F&A, inc Ⓥ), ☎ 702373, 🚢 702708, Slip, BH (28 ton), AC, P, D, FW, ME, M, El, C (8 ton), Sh;
　　Club de Mar R, Bar. Village: few shops.
　　Bueu: essentially a FV hbr, but possible AB outside the E mole, or ⚓ to W of hbr in about 5m. Caution: mussel rafts to NNW and NNE. Hbr ent is lit, Fl G 3s and Fl (2) R 6s. Hr Mr ☎ 320253. P & D (cans), CH, V, R, Bar. Ferry to Isla Ons.
　　Ria de Aldán: Worth exploring in settled weather; beware mussel rafts on W side. Temp'y ⚓s may be found on E side and at head of ría; uncomfortable in fresh N'lies.
　　Isla Ons: Ferry jetty at Almacen. Fair weather ⚓ in 5m off the beach at Melide. Almost uninhabited, but tourist trap in season. Bar, R at Almacen. No landing on Isla Onza.

* Marina lies inside new L-shaped bkwtr which extends S, then E from near original mole hd. Fl(4)R 12s. * Q.R. at hd of new bkwtr 42°23'·24N 08°48'·01W. Small inner hbr is full of FVs; unlit Wtr G top marks shoal. Marina ☎(34) 986 720527; 🚢 720578; 500 AB, D, BH, FW, AC, CN.

25m 75ft

Depths are less than charted in the vicinity of Isla Ons, Isla Onza, Canal de Camoucos, Paso de la Fagilda, Ria de Aldán ~~ents~~ and Ria Pontevedra. New depths range from 4.1m to 8.6m.

RÍA DE VIGO IG 24

Galicia 42°14'·63N 08°43'·35W (Ent to Vigo marina)

CHARTS
AC 1757, 2548, 3633; SC 9240/1, 4165/6; SHOM 5837, 5517

TIDES
Standard Port LISBOA (→); ML 1·96; Zone –0100

Times				Height (metres)			
High Water		Low Water		MHWS	MHWN	MLWN	MLWS
0500	1000	0300	0800	3·8	3·0	1·4	0·5
1700	2200	1500	2000				
Differences VIGO							
+0040	+0100	+0105	+0125	–0·4	–0·3	–0·1	0·0

SHELTER
The ent to ría is protected by Islas Cies with easy appr chans from N and S. Good shelter in Vigo marina, unless a NW'ly scend enters. Narrow ent to marina, marked by PHM/SHM lt twrs, is close E of the most E'ly of 4 conspic blue cranes. The old YC bldg, like a liner's superstructure, is in centre of marina. Cangas on the N shore is a sizable town with FV hbr; ⚓ inside or to the E. Very attractive ⚓s on E side of Islas Cies off Isla del Norte, Isla del Faro and Isla de S. Martin, all of which form a Nature Reserve. Ensenada de San Simón, beyond motorway suspension bridge, is shallow but scenic with ⚓s on W, E and S sides.

NAVIGATION
North Chan WPT 42°16'·62N 08°54'·58W, 309°/129° from/to Cabo del Home front ldg lt, 2·2M. Isla del Norte (Cies) lies SW; the chan is 1·4M wide and clear.
South Chan WPT 42°09'·42N 08°55'·00W, 249°/069° from/to Cabo Estay front ldg lt, 5·0M. Caution: on N side of 1·5M wide chan, Castros de Agoeiro shoal (4·1m) lies 5ca S of Isla Boeiro lt ho. A WCM lt buoy defines SE side of chan and marks Las Serralleiras, rky islets NW of Bayona. Inside the ría beware extensive unlit mussel rafts along the N shore. The docks and city of Vigo line the SE shore.

LIGHTS AND MARKS
Islas Cies are readily identified by their high, rugged bare slopes and white beaches on the E side. Ldg lts/marks for the appr chans are as per the chartlet. The buoyed inner fairway lies between brgs of 066° and 077° on the conspic Hermitage (chapel) tower on Monte de la Guía.

RADIO TELEPHONE
Port Control Vigo Prácticos VHF Ch 14 16; Marina Ch 09.

TELEPHONE (Dial code 986)
Hr Mr 432055, ☎ 434807; ⌗ & Met via marina; British Consul 437133.

FACILITIES
Vigo marina (413 F&A + few Ⓥ), ☎ 224003, ☎ 223514, P & D, Slip, BH (32 ton), AC, FW, ME, El, Ⓔ, CH, C (3 ton), Sh, ▢; **Real Club Náutico** ☎ 432603, R, Bar.
City: all amenities; ✈ ; ✈ (national).

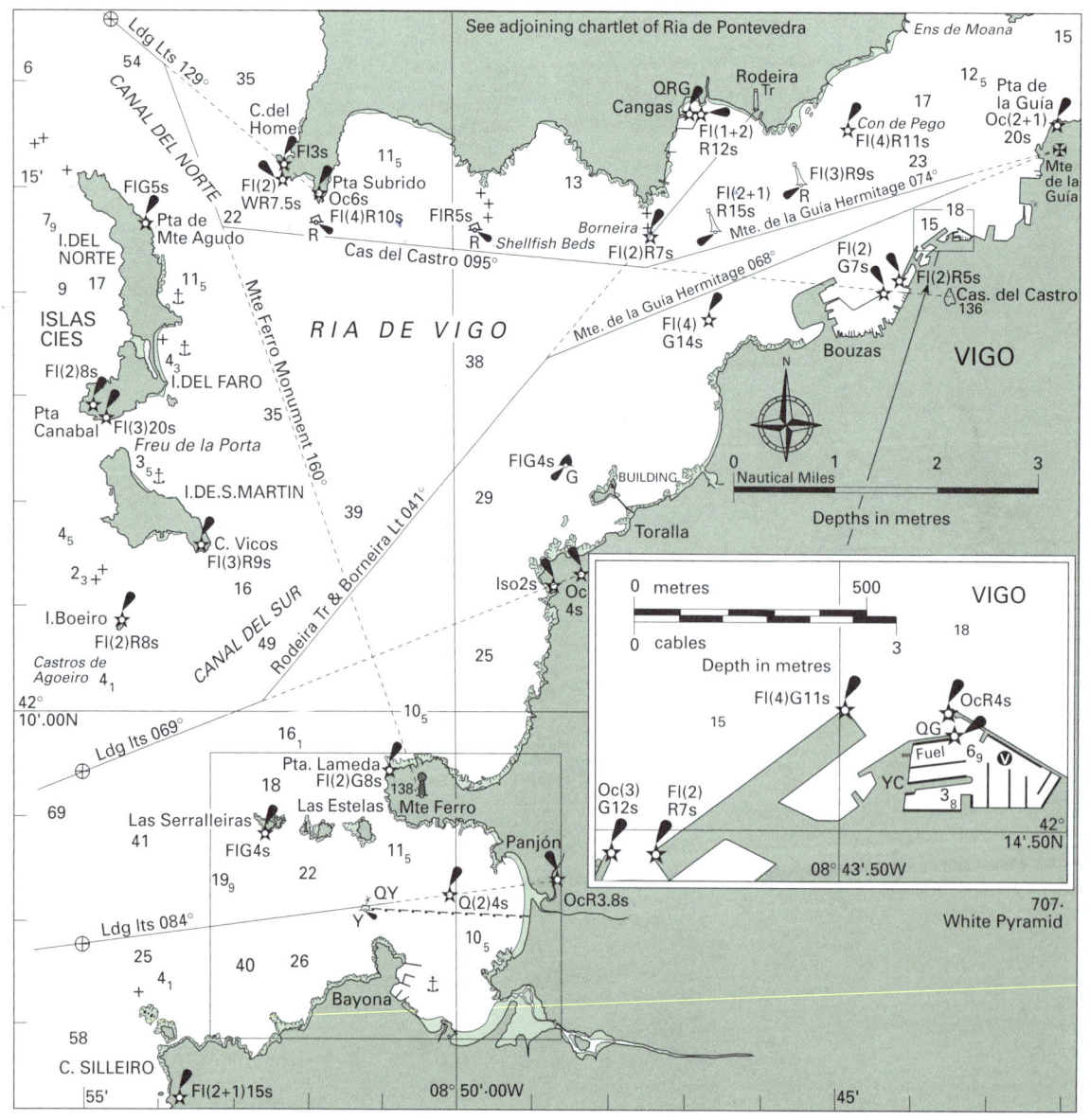

BAYONA   IG 25

Galicia 42°07'·45N 08°50'·55W

CHARTS
 AC 2548, 3633; SC 4167, 9241, 924/5; SHOM 5517, 3007
TIDES
 Standard Port LISBOA (→); ML No data; Zone –0100

Times				Height (metres)			
High Water		Low Water		MHWS	MHWN	MLWN	MLWS
0500	1000	0300	0800	3·8	3·0	1·4	0·5
1700	2200	1500	2000				
Differences BAYONA							
+0035	+0050	+0100	+0115	–0·3	–0·3	–0·1	0·0
LA GUARDIA							
+0040	+0055	+0105	+0120	–0·5	–0·4	–0·2	–0·1

SHELTER
 Excellent; berth in marina, run by YC or on their ⚓s. Or ⚓ E of marina, but clear of fairway to FV jetty. Two disused pontoons on S side of Dique de Abrigo are also available.
NAVIGATION
 WPT 42°07'·80N 08°55'·00W, 264°/084° from/to SPM buoy QY, 2·8M. From S keep 1M off Cabo Silleiro to clear reefs. To the N of ldg line, 4 islets and outliers are marked by SCM and WCM lt buoys. The Canal de la Porta, between Monte Ferro and Las Estelas is a useful shortcut to/from Ría de Vigo, but keep W of 0·9m patch in mid-channel.
LIGHTS AND MARKS
 Ldg lts 084° as chartlet, hard-to-see white conical twrs, the front on a tiny rky islet. N of the bay, a prominent white monument is on Monte Ferro, near Pta Lameda. Other marks conspic by day are: castle walls (parador), white wall of Dique de Abrigo and flag-mast in marina. FV jetty has white crane.
RADIO TELEPHONE
 Monte Real Club de Yates VHF Ch 06, 16.

TELEPHONE (Dial code 986)
 Hr Mr: see marina below; ⌗ & Met via YC; Ⓗ 352011; Police 355027.
FACILITIES
 Marina, run by Monte Real Club de Yates, (200+ some Ⓥ in 6m), ☎ 355576, 📠 355061, 1750Ptas, AB, F&A, M, AC, P, D, FW, Slip, BH (20 ton), C (1·5 ton), ME, El, Ⓔ, CH, SM, Sh, Bar, Ice, R.
 Town: most facilities inc ⌧; ⇌, ✈ Vigo (21 km).

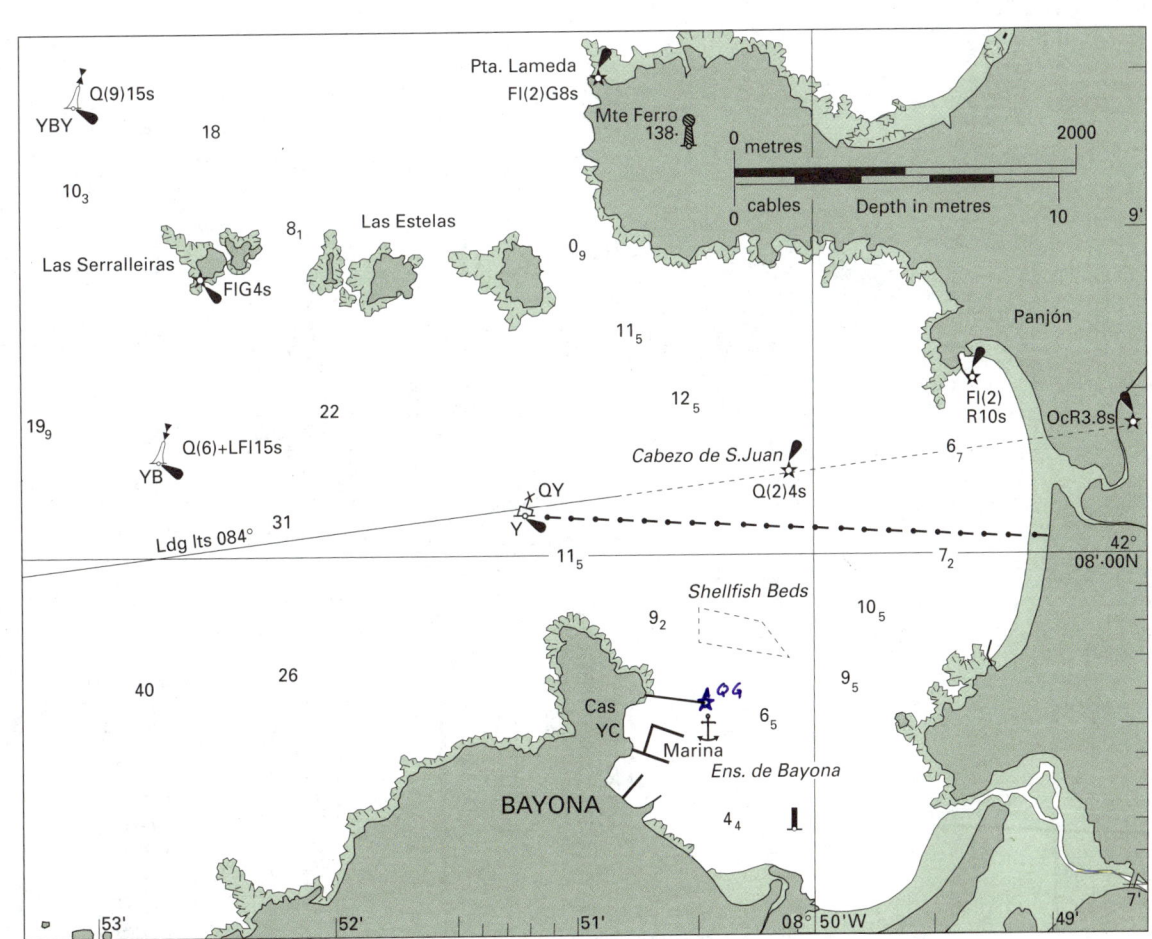

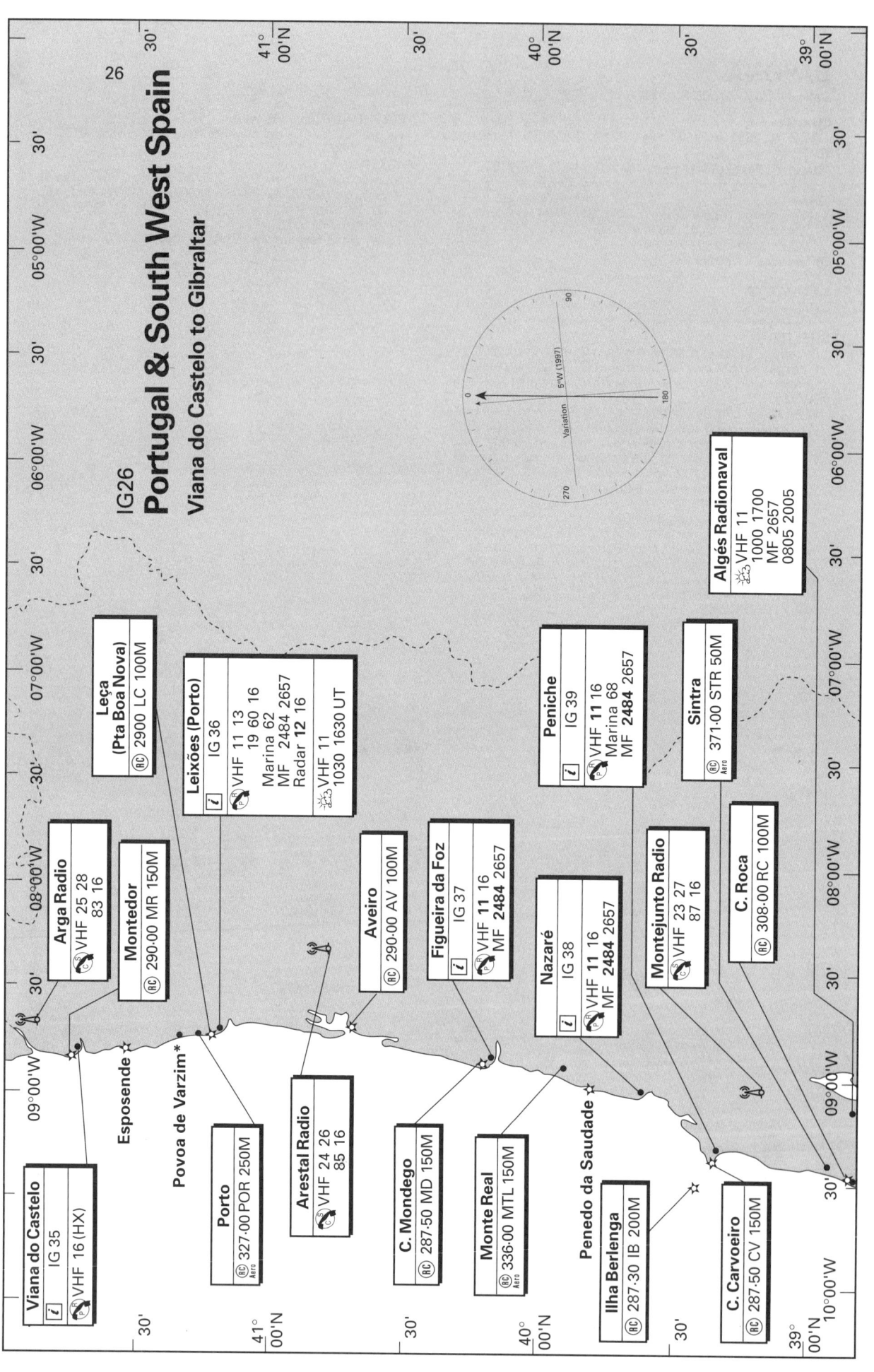

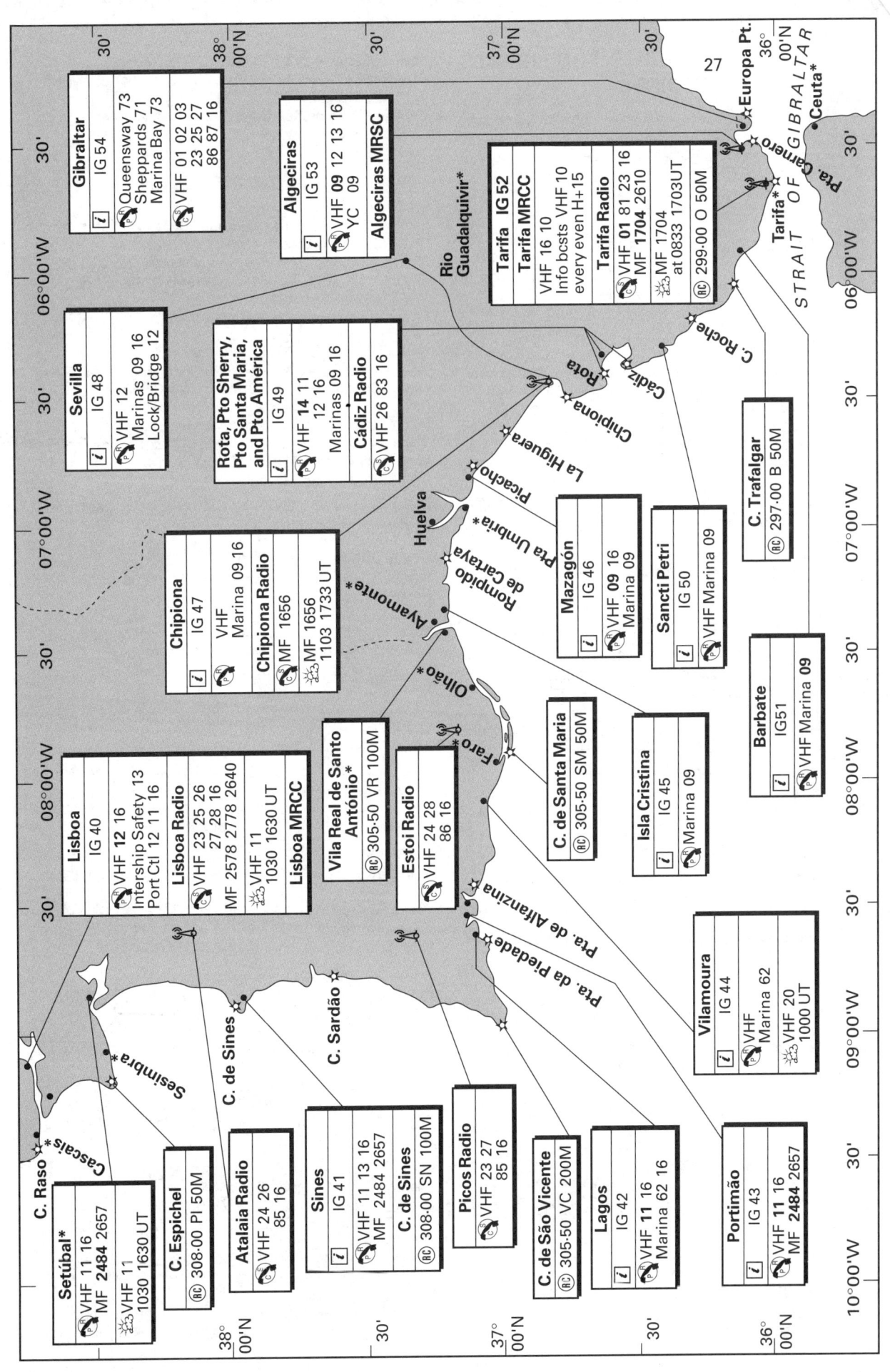

IG 27 COASTAL LIGHTS, FOG SIGNALS AND WAYPOINTS

Abbreviations are as per the Almanac. Lights with a nominal range of 15M or more are in **bold** print, places and features in CAPITALS, and light-vessels, light floats and Lanbys in *CAPITAL ITALICS*. Fog signals are in *italics*. Useful waypoints are underlined. Generally, Spanish waypoints are referenced to ED 50 and Portuguese waypoints to Lisboa datum.

PORTUGAL: RIO MIÑO TO LISBOA

Río Miño ent Forte Ínsua 41°51'·70N 08°52'·40W Fl WRG 4s 15m W12M, R8M, G9M; W ▲ twr.
Montedor 41°45'·00N 08°52'·30W Fl (2) 9·5s 102m **22M**; □ twr and house; *Horn (3) 25s*.

• VIANA DO CASTELO
Ldg lts 012·5°: **Front, Castelo de Santiago SW Battery,** 41°41'·40N 08°50'·26W Iso R 4s 14m **23M**; **rear,** 500m from front, **Senhora da Agonia** Oc R 6s 32m **23M;** both vis 241°-151°.
Molhe Exterior 41°40'·52N 08°50'·57W Fl R 3s 9M; *Horn 30s*.
Esposende (Forte da Barra do Río Cávado) 41°32'·50N 08°47'·40W Fl 5s 20m **21M**; R ○ twr and house; *Horn 20s*.
Regufe 41°22'·45N 08°45'·20W Iso 6s 29m **15M**; R ○ twr.
Póvoa de Varzim Molhe N hd 41°22'·20N 08°46'·20W Fl R 3s 14m 12M; *Siren 40s* (on S hd).

• LEIXÕES
Leça 41°12'·14N 08°42'·63W Fl (3) 14s 57m **28M**; W ○ twr, B bands; RC.
Quebra Mar mole hd 41°10'·44N 08°42'·39W Fl WR 5s 23m W12M, R9M; twr; vis R 001°-180°, W180°-001°; *Horn 20s*.

• RIO DOURO, PORTO
Ent N side, Felgueiras mole hd 41°08'·87N 08°40'·56W, Fl R 5s 16m 9M; vis 265°-134°; 6-sided twr; *Horn 30s*.
Bar ldg lts 078·5°: Front, Cais da Cantareira 41°08'·90N 08°39'·93W Oc R 6s 11m 9M; W I and gallery; rear, 500m from front, Oc R 6s 32m 9M; W lantern on R col, Y bands.
Furadouro 40°52'·30N 08°40'·60W Fl 4s 11m 8M.
Aveiro 40°38'·64N 08°44'·79W Fl (4) 13s 65m **23M**; RC.
Aveiro Molhe N 40°38'·68N 08°45'·72W Fl R 3s 11m 8M; W ▲ twr, R bands; *Horn 15s*.

• FIGUEIRA DA FOZ
C. Mondego 40°11'·35N 08°54'·20W Fl 5s 101m **28M**; W □ twr, and house; RC; *Horn 30s*.
Molhe N hd 40°08'·82N 08°52'·42W Fl R 6s 14m 9M; *Horn 35s*.
Penedo da Saudade 39°45'·90N 09°01'·80W Fl (2) 15s 54m **30M**; □ twr, and house.

• NAZARÉ
SW corner of Morro da Nazaré 39°36'·30N 09°05'·10W Oc 3s 49m 14M; R lantern on wall of fort; vis 282°-192°; *Siren 35s*.
Molhe S hd 39°35'·40N 09°04'·50W L Fl G 5s 14m 8M.

Ponta de Santo António 39°30'·70N 09°08'·50W Iso R 6s 32m 9M; W ○ twr, R bands; *Siren 60s*.
São Martinho do Porto ldg lts 145°: Front 39°30'·15N 09°08'·35W Iso R 1·5s 9m 9M; rear, 127m from front, Oc R 6s 11m 9M.

• ILHA DA BERLENGA/OS FARILHÕES
Farilhão Grande 39°28'·80N 09°32'·70W Fl 5s 99m 13M.
I. da Berlenga 39°24'·95N 09°30'·50W Fl (3) 20s 120m **27M**; W □ twr; RC; *Horn 28s*.

• PENICHE
C. Carvoeiro 39°21'·75N 09°24'·40W Fl (3) R 15s 57m **15M**; W □ twr; RC; *Horn 35s*.
Molhe W hd 39°20'·95N 09°22'·45W Fl R 3s 12m 9M; W ▲ twr, R bands; *Siren 120s*.

Assenta 39°03'·60N 09°24'·80W L Fl 5s 74m 13M.
C. da Roca 38°47'·00N 09°29'·80W Fl (4) 18s 164m **26M**; RC; W □ twr; *Siren 20s*.
C. Raso (Forte de São Brás) 38°42'·47N 09°29'·07W Fl (3) 15s 22m **20M**; R ○ twr; vis 324°-189°; *Horn (2) 60s*.

LISBOA TO SPANISH BORDER

• CASCAIS
Ldg lts 285°: Front, **Pta do Salmodo,** 38°41'·32N 09°25'·19W Oc WR 6s 24m **W18M**, R14M; W □ twr, Bu bands; vis R233°-334°, W334°-098°; *Horn 10s;* rear, **Nossa Senhora de Guia,** 2280m from front, Iso WR 2s **W19M, R16M**, W twr; vis W326°-092°, R278°-292°.
Praia da Ribeira 38°41'·69N 09°25'·14W Oc R 4s 5m 6M.

• LISBOA, RIO TEJO AND APPROACHES
Tejo Fairway buoy 38°36'·23N 09°23'·60W Mo (A) 10s; SWM.
Ldg lts 047° (both vis 039·5°-054·5°): Front, **Gibalta** 38°41'·87N 09°15'·90W Oc R 3s 30m **21M**; W ○ twr; rear, **Esteiro,** 760m from front, Oc R 6s 81m **21M**; W □ twr, R bands; Racon (I).
Mama Sul Iso 6s 154m **21M** (also on 047° ldg line). vis 045·5-048·5 (3)
Forte Bugio 38°39'·54N 09°17'·86W Fl G 5s 27m **21M**; ○ twr on fort; *Horn Mo (B) 30s*.

C. Espichel 38°25'·00N 09°12'·80W Fl 4s 167m **26M**; W 6-sided twr; RC; *Horn 31s*.

• SESIMBRA/SETÚBAL, RIO SADO.
Sesimbra ldg lts 004°: Front 38°26'·70N 09°06'·00W L Fl R 5s 9m 7M; rear, L Fl R 5s 21m 6M.
Setúbal ldg lts 040°: Front, fishing basin E jetty 38°31'·06N 08°53'·87W Oc R 3s 12m 14M; R structure, W stripes; rear, **Azêda**, 1·7M from front, Iso R 6s 71m **17M**, R hut.
No 2 lt bn 38°27'·13N 08°58'·38W Fl (2) R 10s 13m 9M; R hut; Racon (B).
Forte de Outão 38°29'·22N 08°55'·99W Oc R 6s 33m 12M.
Pinheiro da Cruz 38°15'·40N 08°46'·30W Fl 3s 66m 11M.

• SINES
C. de Sines 37°57'·48N 08°52'·75W Fl (2) 15s 50m **26M**; RC.
Mohle Leste hd 37°56'·23N 08°51'·88W L Fl G 8s 16m 6M.
Rio Mira ent Milfontes 37°43'·05N 08°47'·25W Fl 3s 21m 10M.
C. Sardão 37°35'·80N 08°48'·85W Fl (3) 15s 66m **23M**.

• CAPE ST VINCENT
C. de São Vicente 37°01'·30N 08°59'·70W Fl 5s 84m **32M**; W ○ twr; RC; *Horn (2) 30s*.
Pta de Sagres 36°59'·60N 08°56'·90W Iso R 2s 51m 11M.
Baleeira mole hd 37°00'·60N 08°55'·40W Fl WR 4s 12m W14M, R11M; W twr; vis W254°-355°, R355°-254°.

• LAGOS/PORTIMÃO
Pta da Piedade 37°04'·75N 08°40'·10W Fl 7s 49m **20M**.
Lagos mole E hd 37°05'·80N 08°39'·85W Mo (A) G 10s 6M.
Portimão mole E hd 37°06'·58N 08°31'·51W Fl G 5s 8m 3M.

Pta de Alfanzina 37°05'·10N 08°26'·45W Fl (2) 15s 61m **29M**.
Pta da Baleeira 37°04'·75N 08°15'·75W Oc 6s 30m 11M.
Praia da Albufeira E pt of bay 37°05'·10N 08°14'·80W Iso R 3s 21m 8M; W △ twr, R bands.

• VILAMOURA–TAVIRA
Vilamoura 37°04'·38N 08°07'·30W Fl 10s 17m **19M**.
E mole hd 37°04'·12N 08°07'·30W Fl G 4s 13m 5M.
C. de Santa Maria 36°58'·55N 07°51'·78W Fl (4) 17s 49m **25M**; W ○ twr; RC.
Barra Nova mole E hd 36°57'·87N 07°52'·06W Fl G 4s 9m 6M.
Faro. Santo António do Alto 37°01'·24N 07°55'·13W Oc R 6s 63m 6M; Church twr.
Olhão Church belfry 37°01'·65N 07°50'·36W Oc R 4s 20m 6M.
Tavira W mole hd 37°06'·75N 07°37'·00W Fl R 2·5s 8m 7M.

Coastal Lights, Fog Signals and Waypoints

- RÍO GUADIANA
Rio Guadiana chan buoy 37°08'·90N 07°23'·40W Q (3) G 6s; SHM.
W trng wall hd 37°09'·70N 07°23'·90W Fl R 5s 4M.
Vila Real de Santo António 37°11'·10N 07°24'·90W Fl 6·5s 51m **26M**; W ○ twr, B bands; RC.

SOUTH-WEST SPAIN

Ayamonte, E trng wall hd 37°09'·80N 07°23'·50W Fl G 3s 4M.

- ISLA CRISTINA, RÍA DE LA HIGUERITA
W mole hd 37°10'·90N 07°19'·60W VQ (2) R 5s 7m 4M.
Rompido de Cartaya N bank 37°13'·20N 07°07'·60W Fl (2) 10s 41m **24M**; W I, B bands.

- HUELVA, MAZAGÓN
Tanker mooring buoy 37°04'·87N 06°55'·50W Fl (4) Y 20s 8M; SPM; *Siren 30s*.
Picacho 37°08'·18N 06°49'·47W Fl (2+4) 30s 51m **29M**.
Ría de Huelva dique d 37°06'·55N 06°49'·83W Fl (3+1) WR 20s 30m 12/9M; vis W165°-100°, R100°-125°; Racon (K).
Porto Deportivo dique d 37°07'·92N 06°50'·01W QG 7m 2M.
La Higuera 37°00'·60N 06°34'·05W Fl (3) 20s 46m **20M**; twr.

- RÍO GUADALQUIVIR
Chipiona Pta del Perro 36°44'·34N 06°26'·46W Fl 10s 68m **25M**.
New Canal ldg lts 068·9°: Front 36°47'·92N 06°20'·18W Q 27m 10M; R △; rear, 0·6M from front, Iso 4s 60m 10M.
No 1 buoy 36°45'·82N 06°26'·93W L Fl 10s; SWM.
Bajo Salmedina lt bn 36°44'·36N 06°28'·56W Q (9) 15s 9m 5M; WCM; Racon (M).
Bajo El Quemado buoy 36°35'·96N 06°23'·87W Fl (2) R 9s; PHM.

- BAY OF CÁDIZ: ROTA/PUERTO SHERRY
Rota 36°38'·22N 06°20'·76W Aero Alt Fl WG 9s 79m **17M**.
Rota 36°37'·04N 06°21'·36W Oc 4s 33m 13M.
Las Cabezuelas buoy 36°35'·29N 06°19'·88W Q (4) R 10s; PHM.
Rota Mole hd 36°36'·94N 06°20'·96W Fl (3) R 10s 8m 9M.
Naval Base bkwtr 36°36'·72N 06°19'·48W Oc (2) R 6s 16m 4M.
Puerto Sherry bkwtr hd 36°34'·73N 06°15'·17W Oc R 4s 4M.
Santa María bkwtr hd 36°34'·42N 06°14'·87W Fl R 5s 9m 3M.
Castillo de San Sebastián 36°31'·77N 06°18'·87W Fl (2) 10s 38m **25M**; twr on castle; *Horn Mo (N) 20s*.
Cádiz Malecón de San Felipe hd 36°32'·65N 06°16'·68W Fl G 3s 10m 5M; G mast on hut.
Los Cochinos No 1 buoy 36°33'·21N 06°18'·98W Fl G 3s; SHM.
Bajos de San Sebastián buoy 36°31'·38N 06°20'·28W Fl (9) 15s; WCM.

C. Roche 36°17'·80N 06°08'·30W Fl (4) 24s 44m **20M**.
C. Trafalgar 36°11'·00N 06°02'·00W Fl (2+1) 15s 50m **22M**; W △ twr; RC.

- BARBATE
Barbate 36°11'·30N 05°55'·30W Fl (2) WR 7s 22m 10/7M.
Dique de Poniente hd 36°10'·80N 05°55'·45W Fl (2) R 6s 11m 5M.

STRAIT OF GIBRALTAR

Pta de Gracia 36°05'·55N 05°48'·50W Oc (2) 5s 74m 13M.
Pta Paloma 36°03'·95N 05°43'·15W Oc WR 5s 44m 10/7M.

- TARIFA
Tarifa 36°00'·14N 05°36'·50W Fl (3) WR 10s 40m **W26M, R18M**; W twr; vis W113°-089°, R089°-113° RC; Racon (C); *Siren (3) 60s*.
Tarifa Is, E side, 36°00'·31N 05°36'·31W, Fl R 5s 12m 3M, W □ tr, R top.
E mole hd 36°00'·47N 05°36'·13W Fl G 5s 10m 4M.
Pta Carnero 36°04'·71N 05°25'·50W Fl (4) WR 20s 42m **W18M**, R9M; vis W018°-325°, R325°-018°; *Siren Mo (K) 30s*.
Prohib Anch buoy 36°06'·80N 05°24'·67W Fl (3) 10s; ECM.

- ALGECIRAS
App buoy 36°09'·06N 05°24'·99W Mo (A) 4s; SWM.
Dique Norte (Rompeolas) hd 36°08'·69N 05°25'·72W Fl (2) R 6s 10m 8M; R ○ twr, W band.
Muelle de I. Verde 36°07'·68N 05°26'·42W Fl (3) R 5s 6m 2M.

- GIBRALTAR
Aero Lt 36°08'·66N 05°20'·51W Mo (GB) R 10s 405m **30M**.
S mole 'A' hd 36°08'·12N 05°21'·78W Fl 2s 18m **15M**; *Horn 10s*.
N mole 'E' hd 36°08'·98N 05°21'·88W FR 28m 5M.
Europa Pt Victoria twr 36°06'·67N 05°20'·62W Iso 10s 49m **19M** ~~and Oc R 10s 15M. FR 44m 15M; W ○ twr, R band; vis W197°-042°, R042°-067°, Horn 20s.~~ *[handwritten: vis 197°-042° and 067°-125°: same Lt: OcR 10s 15M and FR 44m 15M both vis R042°-067° Horn 20s; W○twr R band 36°06·67N 05°20·62W]*

MOROCCO

C. Spartel 35°47'·55N 05°55'·32W Fl (4) 20s 95m **30M**; Y □ twr; RC; *Dia (4) 90s*.
Monte Dirección (El Charf) 35°46'·10N 05°47'·25W Oc (3) WRG 12s 89m **W16M**, R12M, G11M; vis G140°-174·5°, W174·5°-200°, R200°-225°.
Pta Malabata 35°49'·10N 05°44'·90W Fl 5s 77m **22M**.
Pta Almina 35°53'·95N 05°16'·80W Fl (2) 10s 148m **22M**; W twr; *Siren (2) 45s*.

GLOSARIO PORTUGUÊS   IG 28
See IG 3 for Spanish glossary

A. NAVIGATION — NAVEGACIÓN

English	Português
Beacon (Bn)	baliza
Buoy	bóia
Can (PHM)	cilíndrica
Cone, conical (SHM)	cónica
Isolated danger (IDM)	perigo isolado
Landfall (SWM)	aterragem
Special mark (SPM)	marca especiá
Leading line, transit	enfiamento
Port (side)	bombordo
Starboard	estibordo
Topmark	alvo
Red, (R)	vermelho
White (W)	branco
Black (B)	preto
Yellow (Y)	amarelo
Green (G)	verde
Stripe/band	faixas verticais/horizontais
Lighthouse	farol
Lightship	barco-farol
Fixed (F)	luz fixa
Alternating (Al)	luz alternada
Flashing	luz relâmpagos
Quick flashing (Q)	relâmpagos rápidos
Occulting (Oc)	ocultações
Leading light	farol de enfiamento
Obscured	obscurecido
Whistle	apito
Bell	sino
North (N)	norte
South (S)	sul
East (E)	este
West (W)	oeste
Chart Datum (CD)	zero hidrográfico
High Water (HW)	preia mar *(PM)*
Low Water (LW)	baixa mar *(BM)*
Neaps (np)	águas mortas
Springs (sp)	águas vivas
Slack water, stand	águas paradas

Range	*amplitude de maré*
Tide tables	*Tabela de marés*
Tidal stream atlas	*Atlas de marés*
Flood/ebb stream	*corrente enchente/vasante*
Rate/set (tide)	*força/direcção*
Knot (kn)	*nó*
Height, headroom, clearance	*altura*
Bar	*barra*
Bay	*baía*
Sandhill, dunes	*dunas de areia*
River	*rio*
Mussel beds/rafts	*viveiros*
Strait(s)	*estreito*
Point, headland	*ponta*
Island	*ilha*
Bridge	*ponte*
Conspicuous (conspic)	*conspicuo*
Rock	*rocha*
Wreck	*naufrágio*
Shoal	*baixo*
Reef	*recife*
Anchorage	*fundeadouro*
Breakwater, mole	*quebra-mar, molhe*
Basin	*doca, bacia*

B. FACILITIES — *FACILIDADOS*

Yacht harbour, marina	*doca de recreio*
Harbour Master	*capitanía*
Coastguard (CG)	*policia marítima*
Customs (#)	*alfândega*
Registration number	*número do registo*
Insurance certificate	*certificado de seguro*
Length overall (LOA)	*comprimento*
Beam	*boca*
Draught	*calado*
Mooring buoy	*bólia de atracação*
Dredged	*dragado*
Jetty	*molhe*
Slipway (slip)	*rampa*
Lock	*eclusa*
Lifeboat (LB)	*barco salva-vidas*
Chandlery (CH)	*aprestos*
Crane (C)	*guindaste*
Boatyard (BY)	*estaleiro*
Sailmaker (SM)	*veleira*
Engineer (ME)	*engenheiro*
Boat hoist (BH)	*portico elevador*
Fresh water (FW)	*aguada*
Diesel (D)	*gasóleo*
Petrol (P)	*gasolina*
Paraffin	*petróleo*
Methylated spirits	*alcool metílico*

C. METEOROLOGY — *METEOROLOGIA*

High (anticyclone)	*anticiclone*
Ridge (high)	*crista*
Low (depression)	*depressão*
Trough	*linha de baixa pressão*
Pressure, rise/fall	*pressão, subida/descida*
Front, warm/cold	*frente, quente/fria*
Calm (F0)	*calma*
Light airs (F1)	*aragem*
Light breeze (F2)	*vento fraco*
Gentle breeze (F3)	*vento bonançoso*
Moderate breeze (F4)	*vento moderado*
Fresh breeze (F5)	*vento frêsco*
Strong breeze (F6)	*vento muito frêsco*
Near gale (F7)	*vento forte*
Gale (F8)	*vento muito forte*
Severe gale (F9)	*vento tempestuoso*
Storm (F10)	*temporal*
Squall	*borrasca*
Gust	*rajada*
Drizzle	*chuvisco*
Shower	*aguaceiro*
Rain	*chuva*
Hail	*saraiva*
Thunderstorm	*trovoada*
Cloudy	*nublado*
Visibility, poor; good	*fraca, má; bôa*
Haze	*cerração*
Mist	*neblina*
Fog	*nevoeiro*
Swell	*ondulação*
Choppy	*mareta*
Short/steep (sea state)	*mar cavado*
Rough sea	*mar bravo*
Slight sea	*mar chão*
Overfalls (tide race)	*bailadeiras*
Breakers	*arrebentação*

D. FIRST AID — *PRIMEIROS SOCORROS*

Burn	*queimadura*
Shock	*choque*
Heart attack	*ataque de coração*
Unconscious	*sem sentidos*
Stomach upset	*cólicas*
Dehydration	*desidratação*
Sunburn	*queimadura de sol*
Pain	*dôr*
Fever	*febre*
Toothache	*dôr dos dentes*
Bleeding	*sangrar*
Drown, to	*afogar-se*
Swelling	*inchação*
Poisoning	*envenenamento*
Painkiller	*analgésico*
Antibiotic	*antibiótico*
Splint	*colocar em talas*
Bandage	*ligadura*
Sticking plaster	*adesivo*
Stretcher	*maca*
Dentist	*dentista*
Chemist	*farmácia*

E. ASHORE — *À TERRA*

Bakery	*padaria, pastelaria*
Butcher shop	*açougue*
Post Office	*correio (CTT)*
Stamps	*sellos*
Market	*mercado*
Beach	*praia*
Bus station	*estação de camionetas*
Railway station	*estação de comboios*
Launderette	*lavanderia*
Ironmonger	*ferreiroa*

Radio navigational aids

IG 29 RADIO NAVIGATIONAL AIDS

GPS & Loran: See IG 4. *DGPS beacons are on trial at*
o Carvoeiro 39°21'.53N 09°24'.40W 301.0‡ —
o Espichel 38°24'.83N 09°12'.90W 306.0‡ —

29.1 Marine RDF Beacons

29.1.1 PORTUGAL. Marine RDF beacons operate in 4 groups of 3 beacons. Each group has a common frequency and beacons transmit in sequence, as shown below.

Sequence No	1	2	3	4	5	6
Transmits at H + ... minutes	00	01	02	03	04	05
	06	07	08	09	10	11
	12	13	14	15	16	17
	18	19	20	21	22	23
	24	25	26	27	28	29
	30	31	32	33	34	35
	36	37	38	39	40	41
	42	43	44	45	46	47
	48	49	50	51	52	53
	54	55	56	57	58	59

Montedor Ⓐ   Seq: 1, 4
    41°44'.95N 08°52'.30W   290.00   MR   150M

Leça, Pta Boa Nova Ⓐ   Seq: 2, 5
    41°12'.50N 08°42'.64W   290.00   LC   100M

Aveiro Ⓐ   Seq: 3, 6
    40°38'.47N 08°44'.80W   290.00   AV   100M

C. Mondego Ⓑ   Seq: 1, 2
    40°11'.42N 08°54'.20W   287.50   MD   150M

Ilha Berlenga Ⓑ   Seq: 5, 6
    39°24'.83N 09°30'.50W   287.50   IB   200M

C. Carvoeiro Ⓑ   Seq: 3, 4
    39°21'.53N 09°24'.40W   287.50   CV   150M
This beacon experimentally transmits morse ident 'CV' 3 times per minute on 301.5 kHz

C. Roca Ⓒ   Seq: 1, 2
    38°46'.70N 09°29'.82W   308.00   RC   100M

C. Espichel Ⓒ   Seq: 3, 4
    38°24'.83N 09°12'.90W   308.00   PI   50M

C. de Sines Ⓒ   Seq: 5, 6
    37°57'.48N 08°52'.75W   308.00   SN   100M

C. de São Vicente Ⓓ   Seq: 1, 2
    37°01'.32N 08°59'.70W   305.50   VC   200M

C. de Santa Maria Ⓓ   Seq: 3, 4
    36°58'.38N 07°51'.80W   305.50   SM   50M

Vila Real de Santo António Ⓓ   Seq: 5, 6
    37°11'.28N 07°24'.90W   305.50   VR   100M

29.1.2 SOUTHWEST SPAIN

C. Trafalgar
    36°11'.06N 06°02'.06W   297.00   B   50M

Tarifa
    36°00'.13N 05°36'.47W   299.00   O   50M

Rota
    36°37'.69N 06°22'.77W   303.00   D   80M

29.2 Aeronautical RDF beacons

Porto ⊛
    41°19'.07N 08°41'.98W   327.00   POR   250M

Monte Real ⊛
    39°54'.54N 08°52'.92W   336.00   MTL   150M

Sintra ⊛
    38°52'.92W 09°24'.04W   371.00   STR   50M

Caparica ⊛
    38°38'.62N 09°13'.21W   389.00   CP   250M

Faro ⊛
    37°00'.43N 07°55'.48W   332.00   FAR   50M

Rota ⊛
    ~~36°37'.69N 06°22'.77W   303.00   D   50M~~
    36°38'.62N 06°19'.00W   267.60   AOG   40M

29.3 Racons (Radar beacons)

29.3.1 PORTUGAL

~~R. Tejo fairway buoy~~   38°36'.23N 09°23'.60W   ~~C~~
Esteiro rear ldg lt   38°42'.14N 09°15'.51W   Q
Barreiro 13B buoy (Lisbon)   38°39'.38N 09°05'.75W   Q
Setúbal bn 2   38°27'.13N 08°58'.38W   B

29.3.2 SOUTHWEST SPAIN

Huelva dique head   37°06'.56N 06°49'.85W   K
Bajo Salmedina (Chipiona)   36°44'.36N 06°28'.55W   M
Tarifa lt ho   36°00'.13N 05°36'.47W   C

IG 30 WEATHER SOURCES

30.1 General

Daily forecasts can usually be obtained from YC's or Hr Mr's. The geographic limits of the forecast areas used by UK, French, Spanish and Portuguese authorities do not necessarily coincide, even though similar names are used. Forecasts issued by different nations may also differ markedly in scope and quality.

30.2 BBC Shipping Forecasts

BBC Radio 4 shipping forecasts for Biscay, Finisterre and Trafalgar are broadcast daily on 198 kHz (1515m) at 0048, 05~~3~~5, *1201* ~~1355~~ and ~~1750~~LT* and can probably be received in Portugal, especially at night. Sea Areas Finisterre and Trafalgar are superimposed on Fig. 2 overleaf. ~~*Times are subject to change wef April 1998.~~ *Sea Area Trafalgar is included only in the 0048/et.*

30.3 Radio France Internationale

Radio France Internationale broadcasts a 24h forecast in French for Ouest Portugal (ie from the Portuguese coast to 43°N 11°W to 37°N 11°W) on 11700 kHz at 1145UT. Other frequencies are 6175, 15300, 15530, 17575 kHz.

30.4 Navtex

Stations transmitting in Portugal and W Spain, with identification letter, times (UT) of navigational warnings and forecasts (**shown in bold type**), all in English, are:

Coruña (D) 0030 0430 0830 **1230** 1630 2030
    Weather for Spanish areas 1-6.

Monsanto (R) 0250 0650 1050 1450 1850 2250
    Weather for Zona Norte, Zona Centro and Zona Sul

Tarifa (G) 0100 0500 **0900** 1300 1700 **2100**
    Weather for Spanish areas 6 and 7

30.5 Radio facsimile

Stations transmitting MF/HF SSB radio facsimile weather charts covering Iberia and adjacent sea areas include:

~~**Madrid (ECA7)**~~ ~~Times are UT~~.

Frequencies: 3650, 6918.5, 10250 kHz

Schedule: Surface analysis at 0350 1545
                 Significant wx at 0440 0910 1700 2240
                 Wave analysis at 1105 1215 1730
                 24h wave prognosis at 1215
                 ~~Sea surface temperature at 1715~~

Continued

32 Iberian Guide

Rota (US Navy) (AOK) Times are UT.

Frequencies: 7595 (1800-0600), 9050 (H24), and 10542 kHz (0600-1800)

Schedule: Bcst schedule at 0430 0442 1630 1642
Surface analysis at 0048 1248
24h surface prognosis at 0148 1348
36h surface prognosis at 0824 2024
48h surface prognosis at 0924 2124
72h surface prognosis at 0700 1900
96h European fcst at 0524 1724
Wave heights at 0030 0415 1230 1615
~~US Navy Surface analysis at 1024 2224~~

30.6 Recorded telephone forecasts.

Within Portugal ☎ 0601 123 followed by: 140 ~~for Offshore from R Minho-Porto; and 141 Porto-Lisbon~~. Or 124 for ~~Inshore from R Minho-Porto, and 123 Porto-South~~. For a 9 day general forecast ☎ 0601 123 131. All forecasts are in Portuguese.

In SW Spain ☎ 906 365 373 for offshore bulletins for forecast areas San Vicente, Cádiz, Alborán, Azores and Canaries; also coastal bulletins from Portugal to Gibraltar, and Canaries. All forecasts are in Spanish. The service is only available within Spain and to Autolink-equipped vessels.

(handwritten annotation):
	Offshore	Inshore
N. Border – Lisboa	140	123
Lisboa – C. St. Vincent	141	124
CStV – E. Berdo	142	125

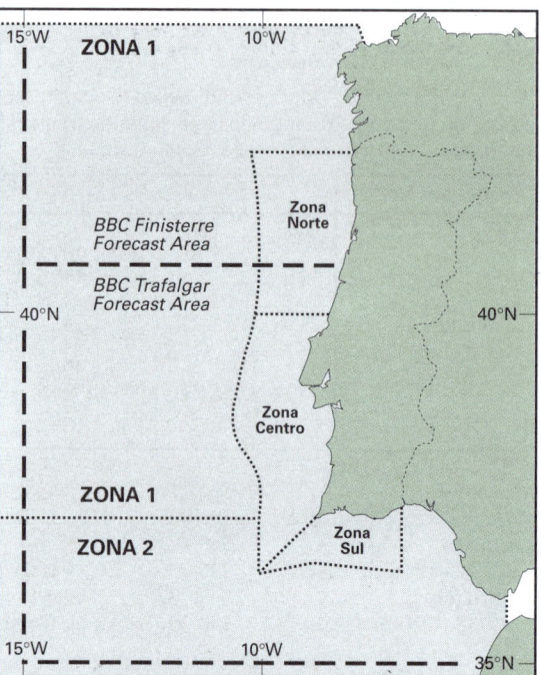

Fig. 2 Portuguese forecast areas
…………… Portuguese Forecast Areas
— — — UK Forecast Areas (BBC)

30.7 Broadcasts of Gale Warnings and Forecasts for Offshore and Coastal Waters

Note: Navigational warnings & traffic lists, as provided by Coast Radio Stations, are on the next page.

PORTUGAL

a. **Radiofusão Portuguesa (National Radio) - Programa 1** broadcasts 24h forecasts for N, Central and S Zones in Portuguese at 1100UT. Transmitters and frequencies are: **Porto** 720 kHz, 96·7 MHz; **Coimbra** 630 kHz; **Lisboa** 666 kHz, 95·7 MHz; **Faro** 720 kHz, 97·6 MHz.

b. The following **Naval Radio Stations** transmit gale warnings and 24h forecasts (also Nav warnings) for N, Central and S Zones (up to 50M offshore) in **English** on 2657 kHz at the times (UT) shown:
Apulia 0735 & 2335; **Monsanto** 0805 & 2005; **Sagres** 0835 & 2035.

SOUTH WEST SPAIN

a. **Chipiona CRS** broadcasts in Spanish gale warnings, synopsis and forecast for Areas 5-8 at 0833 1733UT on 1656 kHz.

b. **Tarifa CRS** broadcasts in Spanish gale warnings, synopsis and forecast for Areas 5-8 at 0803 1703UT on 1704 kHz.

c. **Tarifa MRCC** broadcasts on VHF Ch 10 & 74 at every even H+15 (UT), in Spanish and **English**: Actual wind and vis at Tarifa, followed by forecasts for the Gibraltar Strait, Cádiz Bay and Alboran.

d. **Algeciras MRSC** broadcasts weather messages on VHF Ch 15 & 74 at 0315, 0515, 0715, 1115, 1515, 1915 and 2315UT *in Spanish and English*

GIBRALTAR

a. **Radio Gibraltar** (Gibraltar Broadcasting Corporation) broadcasts in **English** General synopsis, wind direction and strength, visibility and sea state for area 50M radius from Gibraltar. Frequencies are 1458 kHz, 91·3 MHz, 92·6 MHz and 100·5 MHz. Times (UT) are Mon-Fri: 0610 0930 1030 1230 1300 (in Spanish) 1530 1715; Sat: 0930 1030 1230 1300 (in Spanish); Sun: 1030 1230.

b. **British Forces Broadcasting Service Gibraltar** broadcasts in **English** General synopsis, wind direction and strength, visibility and sea state for area 50M radius from Gibraltar, plus times of HW/LW. Frequencies are 93·5 MHz and 97·8 MHz. Times (LT) are Mon-Fri: 0745 0845 1130 1715 2345, and every H+06 (0700-2400); Sat-Sun: 0845 0945 1230 and every H+06 (0700-1000 & 1200-1400).

c. **Gibraltar Coast Radio Station** broadcasts in **English** on request: Gale warnings, 12 hrs Fcst and outlook for a further 12 hrs, from 0000, 0600, 1200 or 1800, for area radius 50M from Gibraltar. Broadcasts are on VHF Ch 01 04 23 25 27 86 and 87.

Coast Radio Stations

IG 31 COMMUNICATIONS

31.1 COAST RADIO STATIONS – PORTUGAL, SOUTH WEST SPAIN AND GIBRALTAR (See also IG 6)

PORTUGAL (Stations are remotely controlled from Lisboa. All monitor Ch 16 H24)

Arga Radio	41°48'N 08°41'W	VHF 25 28 83
Arestal Radio	40°46'N 08°21'W	VHF 24 26 85
Montejunto Radio	39°10'N 09°03'W	VHF 23 27 87
LISBOA RADIO	38°44'N 09°14'W	VHF 23 25 26 27 28

    MF: Transmits (kHz) 2182, 2578, 2640, 2691, 2781, 3607, 2778, 2694   Receives 2182 (H24)
    Traffic lists: 2694 kHz every even H+05.

Atalaia Radio	38°10'N 08°38'W	VHF 24 26 85
Picos Radio	37°18'N 08°39'W	VHF 23 27 85
Estoi Radio	37°10'N 07°50'W	VHF 24 28 86

SOUTH WEST SPAIN (Stations are remotely controlled from Malaga. All monitor Ch 16 H24)

Chipiona Radio   36°42'N 06°25'W   No VHF
    MF: Transmits 1656, 2182 kHz (H24); receives on 2081, 2182 (H24). *Traffic lists:* 1656 kHz every odd H+33 (except 0133 & 2133). *Navigation warnings:* 1656 kHz. Urgent warnings on receipt, after next silence period and at 0003 0403 0803 1203 1603 2003; other warnings at 0803 2003. **English**/Spanish.

Cádiz Radio   36°21'N 06°17'W   VHF 26 83 *(Autolink)*
    Navigation warnings (in Spanish): Ch 26 on receipt, after next silence period, and at 0903 1603.

Tarifa Radio   36°03'N 05°33'W   VHF 81 23 *(Autolink)*
    MF: Transmits kHz **1704** 2182 (H24). Receives 2129 2182 (H24), 2045 2048. *2610 3290 (Autolink).*
    Traffic lists: 1704 kHz. *Navigation warnings:* 1704 kHz on receipt, after next silence period, and at 0033 0433 0833 1233 1633 2033; coastal waters from Cabo Trafalgar to 4°W, in **English** and Spanish.

GIBRALTAR

GIBRALTAR RADIO   36°09'N 05°20'W   VHF 01 02 03 [04] 23 24 25 27 28 86 87. Autolink: 04 (24 28: 2nd/3rd choices, shared with operator) [are used for manual or Autolink calls]. Navigation warnings for area within 50M radius of Gibraltar: Ch 16 in English, on receipt and at 0018 0418 0818 1218 1618 2018UT.

31.2 COASTGUARD STATIONS

MRCC/MRSC

MRCCs and MRSCs primarily handle Distress, Safety and Urgency communications. The Portuguese Navy coordinates SAR in 2 regions, Lisboa and Santa Maria (Azores). In SW Spain and the Gibraltar Strait MRCC Tarifa coordinates SAR. Gibraltar CRS monitors VHF Ch 16, but outside office hrs only on a restricted basis. Digital Selective Calling (DSC) is operational or planned as shown below. Dates can change considerably, so consult Notices to Mariners for latest information. MRCC/MRSCs also broadcast weather and navigational data as shown in IG 30.7. They do not handle link calls.

PORTUGAL

Station	Position	Contact	Channels
LISBON MRCC MMSI 002630100	38°41'N 09°19'W	☎ 1 4416527; 📠 1 4416159	Ch 11 16 (H24) 2182 2657 kHz (H24) DSC Ch 70 (planned '98); 2187·5 kHz (planned '97)
Apúlia MMSI 002630200	41°28'N 08°45'W		Remotely controlled station DSC Ch 70 (planned 1998)
Sagres MMSI 002630400	37°00'N 08°57'W		Remotely controlled station DSC Ch 70 (planned 1998)

SOUTH WEST SPAIN

Station	Position	Contact	Channels
Huelva	37°16'N 06°56'W		DSC Ch 70 (planned 1997)
Cádiz	36°21'N 06°17'W		DSC Ch 70 (planned 1997)
TARIFA MRCC MMSI 002240994	36°01'N 05°35'W	☎ 56 684 740; 📠 56 680 606	Ch 16 10, 2182 kHz (all H24); Call: *Tarifa Traffic* DSC Ch 70, 2187·5 kHz (H24)
Algeciras MRSC MMSI 002241001	36°08'N 05°26'W	☎ 56 585404; 📠 585402	Ch 07, 15, 16; Call: *Algeciras Traffic* DSC Ch 70 (planned 1997)

PASSAGE INFORMATION   IG 32

BIBLIOGRAPHY
The *W coasts of Spain and Portugal Pilot* (Admiralty, NP 67) covers from Cabo Ortegal to Gibraltar. The *Atlantic Spain and Portugal* guide (Imray/RCC, 3rd edition 1995) covers from El Ferrol to Gibraltar. *Guia del Navegante* has fair cover, in English, of SW Spain, but is limited elsewhere. For a Portuguese glossary, see IG 28. Passage information for the coasts of NW and N Spain is in IG 7.

RÍO MIÑO TO CABO CARVOEIRO (AC 3634, 3635)
The Río Miño (*Minho* in Portuguese), 15M S of Cabo Silleiro, forms the N border between Spain and Portugal. The river ent is difficult and best not attempted.

In summer the Portuguese Trades (*Nortada*) are N'ly F4-6 and the Portugal Current runs S at ½-¾kn. Tidal streams appear to set N on the flood and S on the ebb, but are ill documented. In summer gales are rare; coastal fog is common in the mornings. If N-bound, especially if lightly crewed, it is worth making daily passages between about 0400 and 1200 to avoid the stronger winds in the afternoon, which may be increased by a fresh onshore sea breeze.

The 150M long coastline is hilly as far S as Pôrto, then generally low and sandy, backed by pine forests, to Cabo Carvoeiro; there are few prominent features. Coasting yachts should keep about 3M offshore. Viana do Castelo (IG 35) is a commercial and fishing port, with a marina at the NE end. Except for Leixões (IG 36), an industrial port 2M N of the mouth of the R. Douro, other hbrs can be closed in bad weather due to heavy swell breaking on the bar. The R. Douro is not easily entered; but Oporto can be visited by road from Leixões. Aveiro and Figueira da Foz (IG 37) are both exposed to the west; the latter has a marina and can be identified from N or S by the higher ground (257m) of Cabo Mondego. 34M further SSW Nazaré (IG 38) is an artificial fishing hbr and port of refuge with easy ent and a marina. Cabo Carvoeiro, the W tip of Peniche peninsula, looks like an island from afar (do not confuse with Ilha Berlenga). Peniche fishing port and marina (IG 39) on the S side of the peninsula are well sheltered from N'lies, but always prone to SW swell.

Ilha da Berlenga and Os Farilhões, both lit, are respectively 6 and 9·5M NW of Cabo Carvoeiro, 14M W of which is a N/S orientated TSS. The normal coastal route is between Cabo Carvoeiro and Berlenga; this chan is deep, clear and 5M wide, although it appears narrower until opened up.

CABO CARVOEIRO TO CABO SÃO VICENTE (charts 3635, 3636)
The coastline rises steadily to the high (527m) ridge of Sintra, inland of Cabo da Roca, S of which it drops steeply to the low headland of Cabo Raso. A TSS lies 10M W of Cabo da Roca, the most W'ly point of Europe. Cascais, about 3·5M E of Cabo Raso, is a favoured anch (marina under construction) for yachts on passage or not wishing to go 12M further E to Lisboa. From Cascais to the Rio Tejo (Tagus) use the Barra Norte (least depth 5m) which joins the main Barra Sul abeam São Julião lt. In Lisboa (IG 40) there are 5 marinas along the N bank of R. Tejo which is 1M wide and clear. If S-bound from Lisboa, stand on about 1M SW of Forte Bugio lt before altering toward Cabo Espichel.

E from this flattish cape the coast rises to 500m high cliffs nearing the ent to Rio Sado. The fishing hbr of Sesimbra is well sheltered from the N, or anchor in a shallow bay at Arrábida. The port of Setúbal has limited yacht facilities. Unbroken beach stretches 35M from R. Sado to Cabo de Sines, lt ho and 3 conspic chimneys to the E. At Sines (IG 41), a strategically placed commercial port, a small marina has limited facilities. The 56M rky coast to Cabo São Vicente offers no shelter except at Punta da Arrifana (37°17'·5N), a tiny bay protected from N'lies by high cliffs. 5M off Cabo São Vicente a TSS is orientated NW/SE.

CABO SÃO VICENTE TO CAPE TRAFALGAR (AC 89, 90)
There are passage anchs, sheltered from N'lies, at Enseada de Belixe, E of C. São Vicente, and at Enseadas de Sagres and da Baleeira, NE of Pta de Sagres. E of C. São Vicente the summer weather becomes more Mediterranean, ie hotter, clearer and winds more from the NW to SW. Swell may decrease, tidal streams are slight and the current runs SE. Hbrs along the coasts of the Algarve and SW Andalucía are sheltered from all but S'lies (infrequent).

The choice of routes to Gibraltar is between a direct track (Cabo São Vicente to Tarifa is 110°/175M) or a series of coastal legs. There is a good marina at Lagos (IG 42) and pontoons at Portimão (IG 43), 20M and 25M E of C. São Vicente. Vilamoura (IG 44) is a large, established marina about 10M W of Faro. At Cabo de Santa Maria a gap in the low-lying dunes gives access to the lagoons and chans leading to Faro and Olhão, where some peaceful anchs may be found, although yacht facilities are limited. On the Portuguese bank of the Rio Guadiana there is a marina at Vila Real de Santo António.

On the Spanish bank Ayamonte has few facilities. 4M to the E, Isla Cristina (IG 45) is the most W'ly of 7 modern marinas in SW Andalucía (see IG 45-47 and 49-51). 13M further E, Rio de las Piedras requires care on entry but is a peaceful river with yacht facilities at El Rompido. In the apps to Huelva there are pontoons at Punta Umbria and a new marina at Mazagón (IG 46). About 30M SE is the mouth of Rio Guadalquivir (chart 85), navigable 54M to Sevilla (IG 48). Caution: NW of the river mouth large fish havens in shoal waters extend 5M offshore. There is a marina at Chipiona (IG 47) on the SE side of the ent. Around Cádiz Bay (IG 49) there are marinas at Rota, Puerto Sherry, Santa Maria and in the city of Cádiz itself. S of Cádiz keep about 4M offshore on the 20m line to clear inshore banks. Sancti-Petri marina (IG 50) lies on a remote and interesting river. Cabo Trafalgar may be rounded within 100m of the lt ho, inshore of a tidal race, or about 4M off to clear foul ground/shoals lying from SW to NW of the cape. Tunny nets up to 7M offshore are a hazard, Apr-Oct, especially off Barbate.

THE GIBRALTAR STRAIT (charts 142, 1448)
IG 52 chartlet shows orientation of the TSS and timing of tidal streams with notes on surface flow. Hbrs include Barbate (IG 51), Tarifa, Algeciras (IG 53) and Gibraltar (IG 54). North Africa, with the exception of Ceuta, is not covered in this Guide. Yachts bound to/from Gibraltar will usually use the northern ITZ. *Levantes* are E'lies common in the Strait when pressure is high to the N and low to the S. A persistent *Levante* produces the roughest seas. The *Poniente* is a W'ly wind, slightly less common in summer. The *Vendavale* is a moist SW'ly with rain/drizzle.

SPECIAL NOTES FOR PORTUGAL

IG 33

Districts/Provinces: Mainland Portugal is divided into 18 administrative districts. However the names of the former provinces (on the coast: Minho, Douro, Beira Litoral, Estremadura, Alentejo and Algarve) are still used and are shown below the name of each main port.

Charts: There are 2 folios, the new F94 (very few charts; 5 digit chart nos) and the old FA (*antigo*). Portuguese charts (PC) are available from: Instituto Hidrográfico, Rua das Trinas 49, 1296 Lisboa Codex; ☎ (01) 3955119, 📠 3960515.

Time: Portugal keeps UT as Standard time and UT+1 as DST, from the last Sunday in March until the Saturday before the 4th Sunday in October, as in other EU nations. Spain, however, keeps UT+1 as its Standard Time and UT+2 as DST.

Representation: Portuguese Trade and Tourism Office, 22-25A Sackville St, London W1X 1DE; ☎ 0171 494 1441, 📠 0171 494 1868.
British Embassy, Rua de S. Domingo à Lapa 35-37, 1200 Lisbon; ☎ (01) 396 1191, 📠 676768. There are Brit Consuls at Porto & Portimão (qv).

Telephone: To telephone Portugal from UK dial 00-351, followed by the area code, less the initial 0, then the seven or eight digit ☎ number. To call UK from Portugal, dial 00-44, followed by the area code, less the initial 0, then the ☎ number.

Access: There are daily flights in season to/from the UK via Porto, Lisbon and Faro. These airports are quite well connected by bus and some trains to other towns. See also IG 8 for Spanish flights.

Currency: The Escudo (approx 285/£ in Dec 1997) is the unit of currency. Credit cards are widely accepted; cash dispensers are available in most medium-sized towns.

Public Holidays: Jan 1; Feb 11 (Shrove Tues); 10 Apr (Good Friday); April 25 (Liberation Day); May 1 (Labour Day); Jun 6 (Corpus Christi); Jun 10 (Camões Day); Aug 15 (Assumption); Oct 5 (Republic Day); Nov 1 (All Saints Day); Dec 1 (Independence Day), 8 (Immaculate Conception), 25 (Christmas). Also every town has a local *festa*.

Buoyage conforms to the IALA Region A system.

Documents: Portugal, although an EU member, requires to check paperwork. This can be a time-consuming, repetitive and inescapable process; some hbrs have speeded up their procedures. The only palliatives are courtesy, patience and good humour. Organise your papers to include: *Personal* – Passports; crew list, ideally on headed paper with the yacht's rubber stamp, giving DoB, passport nos, where joined/intended departure. Certificate of Competence (Yachtmaster Offshore, ICC/HOCC etc). Radio Operator's certificate. Form E111 (advised for medical treatment).
Yacht – Registration certificate, Part 1 or SSR. Proof of VAT status. Marine insurance. Ship's Radio licence. Itinerary, backed up by ship's log. The *Livrete* (transit log), used to record entry/dep to/from Portugal, is no longer normally issued.

TIMES (UT) OF SUNRISE & SUNSET

		JAN 1	15	FEB 1	15	MAR 1	15	APR 1	15	MAY 1	15	JUN 1	15	JUL 1	15	AUG 1	15	SEP 1	15	OCT 1	15	NOV 1	15	DEC 1	15
LISBON	SR	0755	0753	0743	0728	0709	0648	0622	0601	0539	0525	0514	0511	0515	0524	0538	0550	0605	0618	0632	0646	0703	0719	0735	0747
	SS	1726	1739	1758	1814	1829	1843	1900	1913	1929	1942	1955	2003	2005	2001	1947	1931	1907	1845	1820	1759	1737	1723	1715	1716
GIBRALTAR	SR	0732	0732	0723	0709	0652	0633	0608	0548	0529	0515	0506	0504	0508	0516	0528	0539	0553	0604	0616	0628	0644	0658	0714	0725
	SS	1717	1730	1748	1802	1816	1829	1843	1855	1909	1920	1933	1940	1943	1939	1927	1912	1849	1829	1805	1746	1726	1714	1707	1708

DISTANCE TABLE: COASTS OF NW SPAIN, PORTUGAL AND SW SPAIN

IG 34

Approximate distances in nautical miles are by the most direct route, whilst avoiding dangers and allowing for Traffic Separation Schemes. Places in *italics* are off UK and North West coast of France. See also Distance Table at IG 9.

1. *Longships*	**1**																			
2. *Ushant (Créac'h)*	100	**2**																		
3. La Coruña	418	338	**3**																	
4. Cabo Villano	439	365	43	**4**																
5. Bayona	510	436	114	71	**5**															
6. Viana do Castelo	537	468	141	98	32	**6**														
7. Leixões (Pôrto)	565	491	169	126	63	33	**7**													
8. Nazaré	659	585	263	220	156	127	97	**8**												
9. Cabo Carvoeiro	670	596	274	231	171	143	114	22	**9**											
10. Cabo Raso	710	636	314	271	211	183	154	62	40	**10**										
11. Lisboa (bridge)	686	652	330	287	227	199	170	78	56	16	**11**									
12. Cabo Espichel	692	658	336	293	233	205	176	84	62	22	23	**12**								
13. Sines	726	692	370	327	267	239	210	118	96	54	57	34	**13**							
14. Cabo São Vicente	777	743	421	378	318	290	261	169	147	104	108	85	57	**14**						
15. Lagos	797	763	441	398	338	310	281	189	167	124	128	105	77	20	**15**					
16. Vilamoura	820	786	464	421	361	333	304	212	190	147	151	128	100	43	27	**16**				
17. Cádiz	911	877	555	512	452	424	395	303	281	238	242	219	191	134	120	95	**17**			
18. Cabo Trafalgar	928	894	572	529	469	441	412	320	298	255	259	236	208	151	139	115	28	**18**		
19. Tarifa	952	918	596	553	493	465	436	344	322	279	283	260	232	175	163	139	52	24	**19**	
20. Gibraltar	968	934	612	569	509	481	450	360	338	295	299	276	248	191	179	155	68	40	16	**20**

VIANA DO CASTELO   IG 35

Minho 41°41'·68N 08°49'·21W (Marina ent)

CHARTS
   AC 3254, 3633/4; PC 53, 31, 1; SC 4110, 41B; SHOM 6475
TIDES
   Standard Port LISBOA (→); ML 2·0; Zone 0 (UT)

Times				Height (metres)			
High Water		Low Water		MHWS	MHWN	MLWN	MLWS
0400	0900	0400	0900	3·8	3·0	1·4	0·5
1600	2100	1600	2100				
Differences VIANA DO CASTELO							
−0020	0000	+0010	+0015	−0·3	−0·3	0·0	0·0
ESPOSENDE (41°32'N)							
−0010	0000	+0010	+0015	−0·6	−0·5	−0·1	0·0
PÓVOA DE VARZIM (41°22'N)							
−0020	0000	+0010	+0015	−0·3	−0·3	0·0	0·0

SHELTER
   Excellent in marina (3m), 1·5M up R. Lima, 200m short of low road/rail bridge; possible strong cross current when entering marina. No other yacht berths; no ⚓ in river.
NAVIGATION
   WPT 41°40'·00N 08°50'·40W, 185°/005° from/to No 1 SHM lt buoy, 7½ca. From WPT keep rear ldg lt on brg 005° to enter the chan, dredged 8m, which curves to NE and is marked by 14 lateral buoys, all Fl R 7s or G 7s. The outer mole hd should be given a wide clearance; best water lies further E. Night entry not advised if any swell.
LIGHTS AND MARKS
   Montedor, Fl (2) 9·5s, is 4M NNW. ✠ dome on Monte Santa Luzia is conspic in transit with ldg lts. But ignore front ldg lt, Iso R 4s; its transit 012·5° passes too close to the outer mole hd (Fl R 3s), and leads into shipyard and FV area.
RADIO TELEPHONE
   Call Porto de Viana VHF Ch 16 (0900-1200; 1400-1700LT).
TELEPHONE (Dial code 058)
   Hr Mr 822168; ℻ 823346; Met, via marina; Police 822345.
FACILITIES
   Marina (150+ 50 ⓥ; F&A on D pontoon), 2210 escs, ☎ 820074, 🛒 829503, Slip, AC, P, D, FW, C (56 ton), YC;
   Services: ME, El, Sh, CH, BY, SM, Gaz, Ⓔ.
   Town: V, R, Bar, ✉, Ⓗ, Ⓑ, ⇌ (1km); ✈ Porto (50 km).

ADJACENT HARBOUR

PÓVOA DE VARZIM, 41°22'·20N 08°46'·00W. Tides above. Busy FV hbr with good shelter in all winds, but in heavy swell ent is rough. Marina on S side is near completion.* ⚓ in centre of hbr in 2·5m off fishing piers. Regufe lt ho, Iso 6s 29m 15M, overlooks the hbr; N and S moles: Fl R 3s and L Fl G 6s. Facilities: D, FW, ME, El, Gaz, ✉, Ⓑ, ⇌.

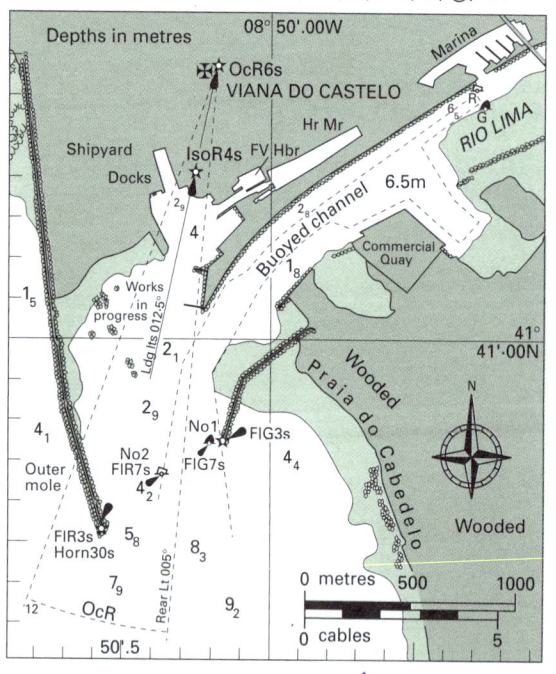

* delayed by financial problems.

LEIXÕES   IG 36

Douro 41°11'·17N 08°42'·18W (Marina ent)

CHARTS
   AC 3254, 3634; PC 58, 21, 22, 16; SC 4210, 42A; SHOM 6475
TIDES
   Standard Port LISBOA (→); ML 2·0; Zone 0 (UT)

Times				Height (metres)			
High Water		Low Water		MHWS	MHWN	MLWN	MLWS
0400	0900	0400	0900	3·8	3·0	1·4	0·5
1600	2100	1600	2100				
Differences LEIXÕES							
−0025	−0010	0000	+0010	−0·3	−0·3	−0·1	0·0
RIVER DOURO ENT							
−0010	+0005	+0015	+0025	−0·6	−0·5	−0·1	0·0
OPORTO							
+0002	+0002	+0040	+0040	−0·5	−0·4	−0·1	+0·1

SHELTER
   Very good in marina (4m), or ⚓ outside in about 4m; no swell once inside N mole. Outer ⚓ is for medium vessels.
NAVIGATION
   WPT 41°10'·20N 08°42'·11W, 170°/350° from/to inner mole hds, 6ca. Ent is simple; keep at least 150m off the N mole hd, due to wreck/sunken bkwtr to S & W of it. Leça lt brg 350° leads into inner hbr, thence steer 015° to marina.
LIGHTS AND MARKS
   Lts as chartlet. Leça lt, Fl (3) 14s 57m 28M, W tr/B bands, is 1·7M N of hbr ent. N of it is oil refinery with many R/W banded chy's. From the S, bldgs of Põrto are conspic.
RADIO TELEPHONE
   Marina Põrto Atlântico VHF Ch 62, 16. Local forecast and gale warnings in Portuguese on Ch 11 at 1030 & 1630UT.
TELEPHONE (Dial code 02)
   Hr Mr 995 1706; ℻ 995 1476; Met 948 4527 (Põrto airport); LB 617 0091; Police 938 0774; Brit Consul 6184789.
FACILITIES
   Marina Põrto Atlântico (200 + 40 ⓥ, F&A), 1880 escs, ☎ 996 4895, 🛒 996 4899, AC, FW, P, D, ME, El, C (6·5 ton), CH, SM, Ⓔ, Sh, Ⓞ; **Clube de Vela Atlântico** ☎ 995 2725; **Clube Naval de Leça** ☎ 995 1700. **Town:** V, Bar, R, Ⓑ, ✉; ⇌ and ✈ Põrto (5km).

ADJACENT HARBOUR

PÕRTO, 41°08'·80N 08°40'·55W. Tides, see above. Difficult river ent, prone to swell and fast current; dangerous in strong W'lies. Limited AB at Cais de Estiva, N bank. Easy to visit by land from Leixões and pre-recce the R. Douro.

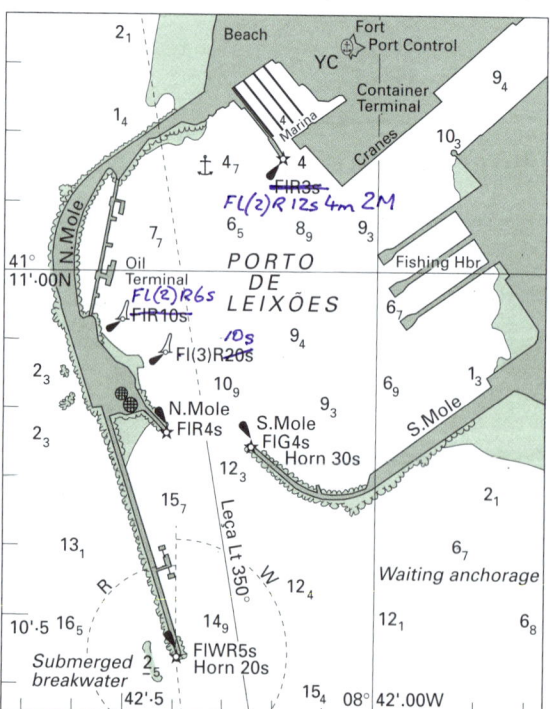

FIGUEIRA DA FOZ   IG 37
Beira Litoral 40°08'·75N 08°52'·40W

CHARTS
AC 3253, 3634, 3635; PC 64, 2, 22, 34; SC 42A; SHOM 7277

TIDES
Standard Port LISBOA (→); ML 2·0; Zone 0 (UT)

Times				Height (metres)			
High Water		Low Water		MHWS	MHWN	MLWN	MLWS
0400	0900	0400	0900	3·8	3·0	1·4	0·5
1600	2100	1600	2100				
Differences BARRA DE AVEIRO							
+0005	+0010	+0010	+0015	−0·3	−0·3	−0·1	0·0
FIGUEIRA DA FOZ							
−0015	0000	+0010	+0020	−0·3	−0·3	−0·1	0·0

SHELTER
Excellent in marina (2·5-3m) on N bank, ¾M from hbr ent. Strong ebb tide can affect pontoons nearest marina ent. Shipyards, repair basins and fishing harbour on S bank.

NAVIGATION
WPT 40°08'·72N 08°52'·70W, 261°/081° from/to hbr ent, 500m. No offshore dangers. Bar at ent is dredged 5m, but shifts/shoals constantly and can be dangerous in swell, especially Nov-Mar with strong S/SW winds. Conditions on the bar may be broadcast on VHF Ch 11 and shown by light/day sigs from Forte de Santa Catarina:
GRG fixed lts (vert) or ● at masthead = Bar closed;
GRG flashing lts (vert) or ● at halfmast = Bar dangerous.

LIGHTS AND MARKS
Lts as chartlet. Cabo Mondego, Fl 5s 101m 28M, is 3M NW of hbr ent. Buarcos, Iso WRG 6s, is on the beach 1·1M N of hbr ent. Ldg lts lead 081·5° past marina ent. White suspension bridge, 1·5M E of ent, is conspic from seaward; as are extensive beaches N and S of hbr ent.

RADIO TELEPHONE
Port *Postradfoz* VHF Ch 11, 16; also 2484kHz.

TELEPHONE (Dial code 33)
Port Authority 22955; Police 28881.

FACILITIES
Marina (150+ 50 Ⓥ), ☎ 22365, ⛴ 23945, AC, FW, Gaz, C, ME, Sh, CH, ⧉; YC ☎ 28019. Town: R, V, Ⓑ, ⛴; ✈ Pôrto.

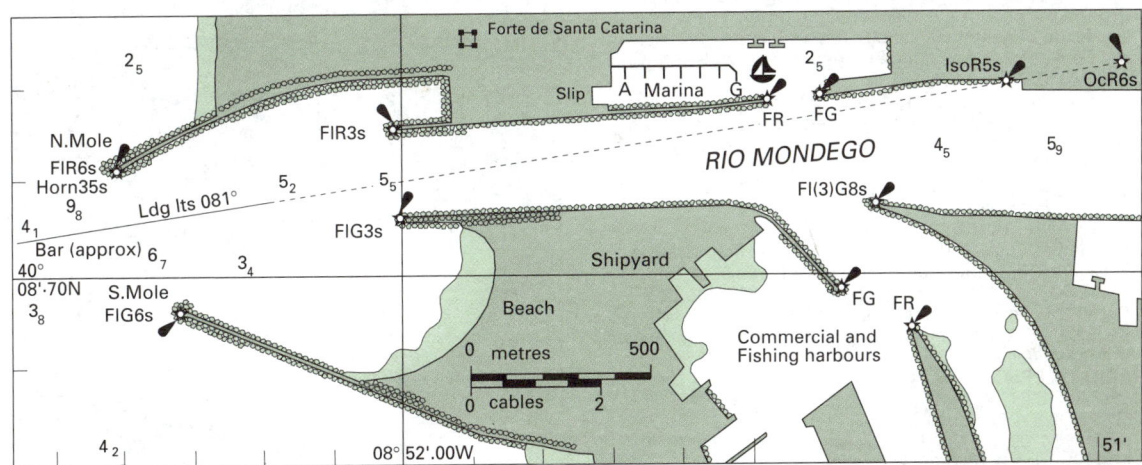

NAZARÉ   IG 38
Estremadura 39°35'·50N 09°04'·57W

CHARTS
AC 3635; PC 65, 34, 35, 22; SC 42B; SHOM 7277

TIDES
Standard Port LISBOA (→); ML 2·0; Zone 0 (UT)

Times				Height (metres)			
High Water		Low Water		MHWS	MHWN	MLWN	MLWS
0400	0900	0400	0900	3·8	3·0	1·4	0·5
1600	2100	1600	2100				
Differences NAZARÉ (Pederneira)							
−0030	−0015	−0005	+0005	−0·5	−0·4	0·0	+0·2

SHELTER
All-weather access to this man-made fishing hbr. Marina is in the SW corner of the inner hbr; visitors should berth on the hammerheads or outer pontoons. The pontoons in the NE corner are for smaller local craft.

NAVIGATION
WPT 39°35'·60N 09°04'·70W, 310°/130° from/to hbr ent, 290m. Due to an underwater canyon there are depths of 150m and 452m 4ca and 5M offshore respectively.

LIGHTS AND MARKS
Hbr Lts as chartlet. Pontal da Nazaré, Oc 3s 49m 14M, siren 35s, is 1M NNW of hbr ent on a low headland which should be cleared by at least 2ca due to offlying rks.

RADIO TELEPHONE
VHF Ch 11, 16 (HO).

TELEPHONE (Dial code 062)
Hr Mr 561255; ⚓ & Met via Hr Mr; Ⓗ 561116.

FACILITIES
Marina (41 + 12 visitors), 3·5m, ☎ 561401, ⛴ 561402, AC, FW, Slip, D & P in NE corner, ME, C, Sh, ⧉; **Club Naval de Nazaré. Town:** (1½M to the N), V, R, Bar, Ⓑ, ✉; ⛴ Valado dos Frades (7km); ✈ Lisbon (125km).

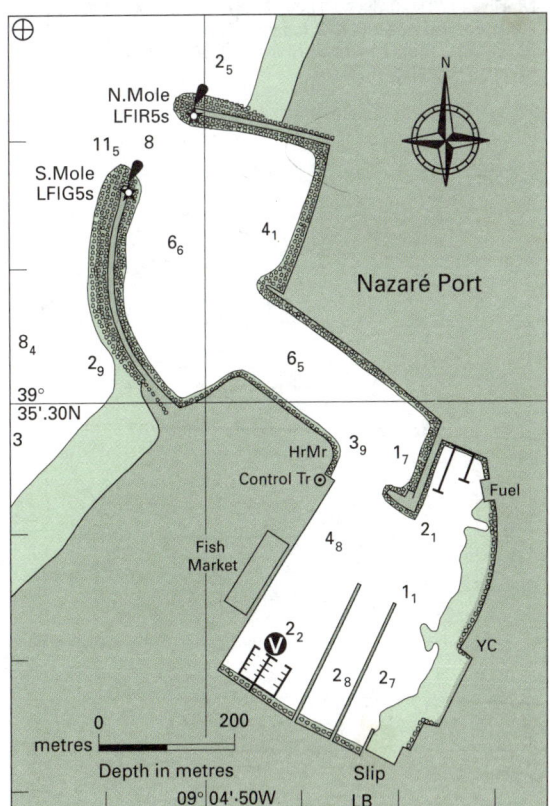

PENICHE

IG 39

Estremadura 39°20'·80N 09°22'·43W

CHARTS
AC 3635; PC 69, 36, 35, 22; SC 42B; SHOM 7277

TIDES
Standard Port LISBOA (→); ML 2·0; Zone 0 (UT)

Times				Height (metres)			
High Water		Low Water		MHWS	MHWN	MLWN	MLWS
0400	0900	0400	0900	3·8	3·0	1·4	0·5
1600	2100	1600	2100				
Differences PENICHE							
−0035	−0015	−0005	0000	−0·3	−0·3	−0·1	+0·1
ERICEIRA (38°58'N)							
−0040	−0025	−0010	−0010	−0·4	−0·3	−0·1	+0·1

SHELTER
A good hbr with all-weather access, but SW swell may work in. Inside W mole berth on outer pontoon of small marina (2·4-3·5m) which is prone to FV wash. Possible moorings or ⚓ in SE part of hbr; holding is good, but ground may be foul. 3kn speed limit in hbr. The ⚓ in 4m on N side of peninsula is open to swell, even in S'lies. On the SE side of Ilha Berlenga there is ⚓ below the lt ho.

NAVIGATION
WPT 39°20'·50N 09°22'·45W, 180°/000° from/to W mole hd, 5ca. The 5M wide chan between Cabo Carvoeiro and Ilha da Berlenga is often rough. Os Farilhões lies 4·5M further to the NW. The E edge of the N/S TSS is 14M W of Cabo Carvoeiro; the ITZ embraces Peniche and the offshore islands.

LIGHTS AND MARKS
Ilha da Berlenga, Fl (3) 20s 120m 27M. Farilhão Grande, Fl 5s 99m 13M. Cabo Carvoeiro, Fl (3) R 15s 57m 15M, is 2M WNW of hbr ent and steep-to apart from Nau dos Carvos, a conspic high rk close-in. Other lts as chartlet. Peniche looks like an island from afar, with conspic water tr in town centre.

RADIO TELEPHONE
VHF Ch 16 (H24), 11, 68.

TELEPHONE (Dial code 062)
Hr Mr 784109, 📠 784767; # & Met via Hr Mr; Ⓗ 781702; Emergency 115 or 789666; LB 789629.

FACILITIES
Marina (120+ 20 Ⓥ), ☎ 784109, 📠 784767, AC, FW; **Clube Náutico; Services:** P & D (cans or tanker), CH, BY, ME, El, Ⓔ, C, Gaz, Slip. **Town:** V, R, Bar, Ⓑ, ✉; ✈ Obidos (30km), ✈ Lisbon (80km).

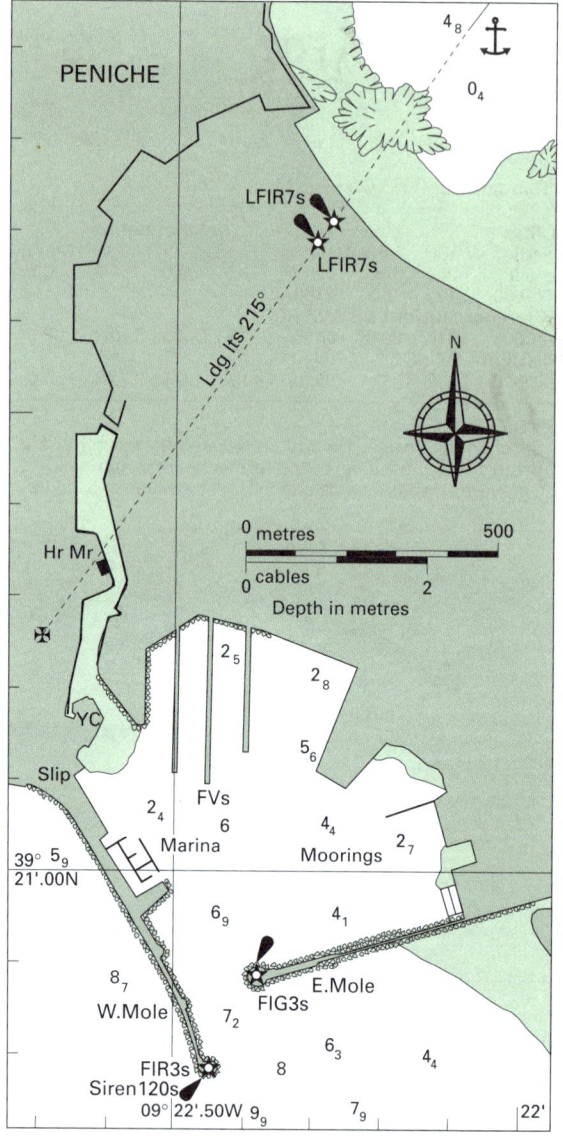

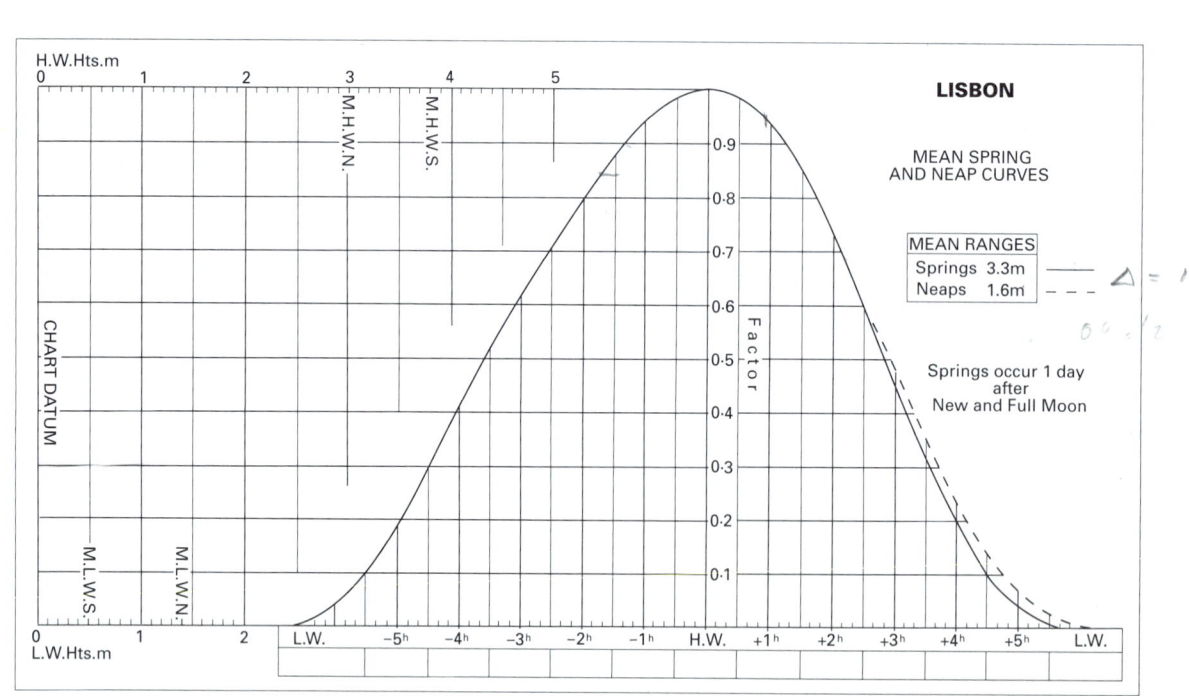

LISBON

MEAN SPRING AND NEAP CURVES

MEAN RANGES
Springs 3.3m
Neaps 1.6m

Springs occur 1 day after New and Full Moon

LISBOA/CASCAIS IG 40

Estremadura 38°41'.30N 09°10'.55W (Ponte 25 de Abril)

CHARTS
AC 3263/4, 3635; PC 45, 46, 48, 263 03/05/06/07; SC 4310, 42B; SHOM 6294/5

Iberian Guide

TIDES
ML 2·0; Zone 0 (UT). Lisboa is a Standard Port for which daily predictions are given below.

Standard Port LISBOA (→)

Times				Height (metres)			
High Water		Low Water		MHWS	MHWN	MLWN	MLWS
0400	0900	0400	0900	3·8	3·0	1·4	0·5
1600	2100	1600	2100				
Differences CASCAIS							
−0040	−0025	−0015	−0010	−0·3	−0·3	+0·1	+0·2
PACO DE ARCOS (5·5M W of Tejo bridge)							
−0020	−0030	−0005	−0005	−0·4	−0·4	−0·1	0·0
SESIMBRA							
−0045	−0030	−0020	−0010	−0·4	−0·4	0·0	+0·1
SETUBAL							
−0020	−0015	−0005	+0005	−0·4	−0·3	−0·1	0·0

SHELTER/FACILITIES
CASCAIS: ⚓ in 3 – 5m, sheltered from prevailing N'lies, but open to S. FW & D from YC. Usual amenities in town; ⛴ to Lisbon. Note: Marina, being built below the citadel walls, should open end May 1999; 660 berths, no visitors.
LISBOA: The 5 marinas, Ⓐ to Ⓔ on chartlet below, are:
Ⓐ. **Doca de Bom Successo**, close E of the ornate Torre de Belem (conspic) and red Pilots bldg. Almost full of local yachts. 100 F&A, 3m, ☎ 301 3027, AC, FW, C, CH, R, Slip, D & P (cans); ⛴, bus to Lisbon.
Ⓑ. **Doca de Belém**, 150m E of Monument to the Discoveries (conspic). Almost full of local yachts; 200 F&A, 3m, ☎ 363 1246, AC, FW, D (hose), BH, C, R; ⛴, bus to Lisbon.
Ⓒ. **Doca de Santo Amaro**, almost below the suspension bridge (road traffic "hums" overhead, plus night life noise). 330 AB (few ⓥ), 4m, ☎ 392 2011/2, 📠 392 2038, AC, FW, R, Bar, V; ⛴, bus 2M to city centre.
Ⓓ. **Doca de Alcântara**, ent is E of the container/cruise ship terminal; chan doubles back W, via swing bridge, toward marina at W end of dock. ~~Swing bridge opens 0815, 0915~~, 1015, 1115, (1215), 1315, ~~1515, 1615, 1800 (1815), 1900, 2000 (2015)~~ & 2200; then stays open (2300) ~~2330~~ to 0730; call *Eclusa* VHF 12. Underlining = only Mon-Fri. Brackets = only Sat, ~~Sun, hols~~. Waiting pontoon is close E. Marina: 180 AB inc ⓥ (best for foreign yachts), 8m, ☎ 392 2011, 📠 392 2048, security fencing, AC, FW, R, Bar, V; ⛴; bus 2M to city centre.

[handwritten: Bridge is reported to be permanently open.]

Ⓔ. **Doca do Terreiro do Trigo** (also known as *APORVELA*, Portuguese STA), approx 700m ENE of the ferry terminal, beyond naval Doca de Marinha. ☎ 887 6854, 📠 8873885, AB inc ⓥ in 1·5m (prone to silting), D, FW, security fence; in run-down area, but close to Alfama (old quarter).
New marina at **Doca dos Olivais** (38°45'·8N 09°05'·5W, off chartlet) may open May 1998 for Expo 98; 500 berths but not for visitors, who may ⚓ at Seixal (38°38'·6N 09°06'·4W) and use ferry. Or consider coach travel from marinas at IG 35, 36, 37, 42 or 44.

City: All amenities; ⛴, ✈. Charts from the Portuguese Hydrographic Office, Rua das Trinas 49, 1296 Lisboa; ☎ 395 5119, 📠 396 0515. ACA: Garraio & Ca Lda, Avenida 24 de Julho 2-1, 1200 Lisboa; ☎ 347 3081, 📠 342 8950.

NAVIGATION
From N/NW keep about 5ca off Cabo da Roca and Cabo Raso. A magnetic anomaly area S of Cabo Raso and W of Guia lt ho can alter local variation by +5° to −3°.
CASCAIS WPT 38°41'·00N 09°24'·73W, 156°/336° from/to Citadel, 6ca. This brg 336° in transit with Peninha (5M NW of Cascais; a rounded 485m summit) also clears the shoals off the ent to R. Tejo for vessels approaching from Cabo Espichel.
LISBOA WPT 38°36'·23N 09°23'·60W, ~~Tejo SWM buoy, Mo (A) 10s~~, 227°/047° from/to betwixt Fort Bugio and Forte de S. Juliao, 5·4M. This appr is the main DW chan, Barra Sul, into R. Tejo proper; shoals lie on either side and break in bad weather. Barra Norte, a lesser chan carrying 5·2m, is a short cut for yachts between Cascais and R. Tejo in fair weather; it is orientated 285°/105° on ldg lts at Forte de Sta Marta and Guia, passing close S of Forte de S. Juliao. Tidal streams run hard (2-3kn) in the river and the ebb sets towards the shoals E of Forte Bugio. Cross-current at the ent to marinas requires care.

LIGHTS AND MARKS
As chartlets and in text.

RADIO TELEPHONE
Call *Porto de Lisboa* (Port authority) VHF Ch **12** 16; work Ch 11. Local forecast and gale warnings are broadcast in Portuguese on Ch 11 at 1030 & 1630UT.

TELEPHONE (Dial code 01)
Hr Mr 608101; ⊞ & Met, via marinas (qv).

OTHER HARBOURS EAST OF CABO ESPICHEL

SESIMBRA 38°26'·20N 09°06'·40W, 5M E of Cabo Espichel. AC 3635, 3636; PC 79; SC42B; SHOM 7238. Tides, in IG 40. A substantial outer bkwtr encloses the FV hbr; yachts may find free AB in an inner basin at NW corner; or ⚓ in about 5m to N of ent and clear of fairway. Caution: strong N'lies blow down from high ground late pm/evening. Lts: ldg lts 004°, both L Fl R 5s 9/21m 7/6M. Outer bkwtr hd, Fl R 3s 8M, W tr, R bands; root, Oc 5s 34m 14M, R ○ tr (Forte de Cavalo). Facilities: FW; tourist amenities.

SETUBAL 38°26'·48N 08°58'·75W; this is chan ent, 1200m from No 2 bn, Fl (2)R 10s, and 6M SW of docks. Tides, see above. AC 3270, 3635/6; PC 82, 263 08/09; SC42B. Commercial hbr with small Doca de Recreio, usually full. Temporary marina reported. Ldg lts 040°: front, Oc R 3s; rear, Iso R 6s. R. Sado is shoal on both sides until past Forte de Outão, Oc R 6s. *[handwritten: Marina operational 1998 W end of Doca Commercial]*
Arrábida (38°28'·5N 08°58'·75W, 1·4M N of ent to R. Sado): popular but shallow ⚓, sheltered from N by high ground.

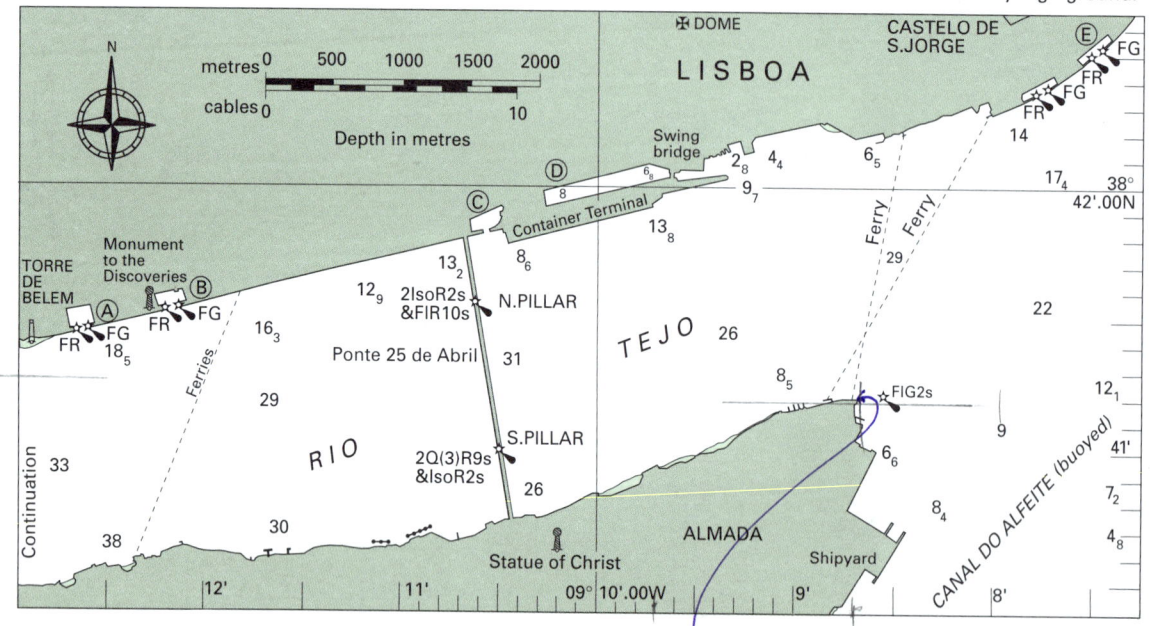

[handwritten: Beware 4 large uncharted stone blocks on the river bed near ferry terminals]

TIME ZONE (UT)
For Summer Time add ONE hour in non-shaded areas

PORTUGAL – LISBON
LAT 38°42′N   LONG 9°08′W

TIMES AND HEIGHTS OF HIGH AND LOW WATERS

YEAR **1998**

JANUARY

Time	m	Time	m
1 0448 / 1028 / TH 1712 / 2242	4.0 / 0.5 / 3.8 / 0.7	**16** 0516 / 1059 / F 1731 / 2305	3.7 / 0.9 / 3.5 / 0.9
2 0532 / 1113 / F 1758 / 2327	4.0 / 0.6 / 3.7 / 0.8	**17** 0549 / 1133 / SA 1804 / 2341	3.6 / 1.0 / 3.4 / 1.1
3 0619 / 1201 / SA 1847	3.9 / 0.7 / 3.6	**18** 0623 / 1210 / SU 1841	3.5 / 1.2 / 3.2
4 0018 / 0710 / SU 1254 / 1941	1.0 / 3.8 / 0.9 / 3.5	**19** 0021 / 0701 / M 1252 / 1924	1.3 / 3.3 / 1.4 / 3.1
5 0116 / 0806 / M 1355 / 2044	1.1 / 3.6 / 1.1 / 3.3	**20** 0109 / 0747 / TU 1344 / 2019	1.5 / 3.1 / 1.5 / 3.0
6 0224 / 0912 / TU 1505 / 2156	1.3 / 3.5 / 1.2 / 3.3	**21** 0212 / 0846 / W 1452 / 2128	1.6 / 3.0 / 1.6 / 2.9
7 0340 / 1024 / W 1619 / 2308	1.4 / 3.4 / 1.2 / 3.4	**22** 0329 / 0959 / TH 1608 / 2243	1.7 / 2.9 / 1.6 / 3.0
8 0456 / 1136 / TH 1728	1.3 / 3.4 / 1.1	**23** 0445 / 1113 / F 1714 / 2349	1.6 / 3.0 / 1.5 / 3.1
9 0014 / 0602 / F 1240 / 1827	3.5 / 1.1 / 3.5 / 1.0	**24** 0546 / 1217 / SA 1808	1.4 / 3.1 / 1.3
10 0111 / 0659 / SA 1337 / 1918	3.7 / 1.0 / 3.6 / 0.9	**25** 0044 / 0638 / SU 1311 / 1856	3.3 / 1.2 / 3.3 / 1.1
11 0202 / 0749 / SU 1426 / 2003	3.8 / 0.8 / 3.7 / 0.8	**26** 0134 / 0724 / M 1401 / 1940	3.5 / 0.9 / 3.5 / 0.9
12 0248 / 0834 / M 1510 / ○ 2044	3.9 / 0.7 / 3.7 / 0.7	**27** 0221 / 0808 / TU 1447 / 2023	3.8 / 0.7 / 3.7 / 0.6
13 0329 / 0914 / TU 1549 / 2122	3.9 / 0.7 / 3.7 / 0.7	**28** 0307 / 0851 / W 1531 / ● 2105	4.0 / 0.5 / 3.8 / 0.5
14 0407 / 0950 / W 1625 / 2157	3.9 / 0.7 / 3.6 / 0.8	**29** 0351 / 0933 / TH 1614 / 2147	4.1 / 0.3 / 4.0 / 0.4
15 0443 / 1025 / TH 1659 / 2231	3.8 / 0.8 / 3.6 / 0.8	**30** 0434 / 1015 / F 1657 / 2229	4.2 / 0.3 / 4.0 / 0.4
		31 0518 / 1058 / SA 1741 / 2313	4.2 / 0.4 / 3.9 / 0.6

FEBRUARY

Time	m	Time	m
1 0602 / 1142 / SU 1827	4.1 / 0.5 / 3.8	**16** 0550 / 1132 / M 1804 / 2343	3.5 / 1.0 / 3.4 / 1.1
2 0000 / 0650 / M 1231 / 1916	0.7 / 3.9 / 0.8 / 3.6	**17** 0623 / 1206 / TU 1841	3.4 / 1.2 / 3.2
3 0052 / 0742 / TU 1326 / 2014	1.0 / 3.7 / 1.0 / 3.4	**18** 0022 / 0702 / W 1247 / 1925	1.3 / 3.2 / 1.4 / 3.1
4 0155 / 0844 / W 1433 / 2123	1.2 / 3.4 / 1.3 / 3.2	**19** 0111 / 0751 / TH 1341 / 2024	1.5 / 3.0 / 1.6 / 2.9
5 0312 / 0959 / TH 1551 / 2243	1.4 / 3.2 / 1.4 / 3.2	**20** 0219 / 0859 / F 1459 / 2143	1.6 / 2.9 / 1.6 / 2.9
6 0438 / 1119 / F 1710 / 2358	1.4 / 3.2 / 1.3 / 3.3	**21** 0348 / 1025 / SA 1626 / 2305	1.6 / 2.8 / 1.6 / 3.1
7 0554 / 1230 / SA 1816	1.3 / 3.3 / 1.2	**22** 0508 / 1144 / SU 1735	1.4 / 3.0 / 1.4
8 0100 / 0653 / SU 1327 / 1908	3.5 / 1.1 / 3.4 / 1.0	**23** 0013 / 0610 / M 1247 / 1831	3.2 / 1.2 / 3.2 / 1.1
9 0151 / 0741 / M 1414 / 1951	3.6 / 0.9 / 3.5 / 0.9	**24** 0110 / 0702 / TU 1340 / 1919	3.5 / 0.9 / 3.5 / 0.9
10 0235 / 0822 / TU 1455 / 2029	3.8 / 0.8 / 3.6 / 0.8	**25** 0201 / 0748 / W 1428 / 2004	3.8 / 0.6 / 3.7 / 0.6
11 0313 / 0857 / W 1530 / ○ 2104	3.8 / 0.7 / 3.6 / 0.7	**26** 0248 / 0830 / TH 1513 / ● 2048	4.0 / 0.3 / 3.9 / 0.3
12 0348 / 0930 / TH 1603 / 2136	3.8 / 0.7 / 3.6 / 0.7	**27** 0332 / 0915 / F 1556 / 2130	4.2 / 0.2 / 4.1 / 0.2
13 0420 / 1000 / F 1633 / 2207	3.8 / 0.7 / 3.6 / 0.7	**28** 0416 / 0956 / SA 1638 / 2212	4.3 / 0.1 / 4.1 / 0.3
14 0450 / 1030 / SA 1702 / 2238	3.8 / 0.7 / 3.6 / 0.7		
15 0519 / 1100 / SU 1732 / 2309	3.7 / 0.9 / 3.5 / 0.9		

MARCH

Time	m	Time	m
1 0459 / 1038 / SU 1721 / 2255	4.3 / 0.2 / 4.1 / 0.4	**16** 0452 / 1030 / M 1704 / 2242	3.7 / 0.8 / 3.6 / 0.8
2 0543 / 1122 / M 1805 / 2341	4.2 / 0.4 / 3.9 / 0.6	**17** 0522 / 1059 / TU 1736 / 2314	3.5 / 0.9 / 3.5 / 0.9
3 0629 / 1207 / TU 1852	3.9 / 0.7 / 3.7	**18** 0554 / 1131 / W 1810 / 2350	3.4 / 1.1 / 3.3 / 1.1
4 0031 / 0719 / W 1259 / 1946	0.9 / 3.6 / 1.0 / 3.4	**19** 0631 / 1208 / TH 1850	3.2 / 1.2 / 3.2
5 0132 / 0819 / TH 1403 / 2054	1.1 / 3.3 / 1.3 / 3.2	**20** 0035 / 0716 / F 1256 / 1943	1.3 / 3.0 / 1.4 / 3.0
6 0250 / 0936 / F 1526 / 2218	1.4 / 3.1 / 1.5 / 3.1	**21** 0135 / 0819 / SA 1406 / 2056	1.5 / 2.8 / 1.6 / 2.9
7 0422 / 1104 / SA 1653 / 2341	1.4 / 3.0 / 1.5 / 3.2	**22** 0301 / 0946 / SU 1540 / 2223	1.5 / 2.8 / 1.6 / 3.0
8 0543 / 1218 / SU 1802	1.3 / 3.1 / 1.3	**23** 0431 / 1114 / M 1702 / 2341	1.4 / 2.9 / 1.4 / 3.2
9 0044 / 0641 / M 1313 / 1853	3.3 / 1.1 / 3.3 / 1.1	**24** 0540 / 1222 / TU 1804	1.1 / 3.2 / 1.1
10 0134 / 0725 / TU 1356 / 1934	3.5 / 1.0 / 3.4 / 0.9	**25** 0043 / 0636 / W 1317 / 1856	3.5 / 0.8 / 3.5 / 0.7
11 0215 / 0802 / W 1433 / 2009	3.6 / 0.8 / 3.5 / 0.8	**26** 0137 / 0724 / TH 1406 / 1943	3.8 / 0.5 / 3.8 / 0.5
12 0250 / 0834 / TH 1506 / 2041	3.7 / 0.7 / 3.6 / 0.7	**27** 0225 / 0810 / F 1451 / 2027	4.0 / 0.2 / 4.0 / 0.3
13 0323 / 0904 / F 1537 / ○ 2112	3.8 / 0.6 / 3.7 / 0.7	**28** 0311 / 0853 / SA 1535 / ● 2111	4.2 / 0.1 / 4.1 / 0.3
14 0353 / 0933 / SA 1606 / 2142	3.8 / 0.6 / 3.7 / 0.6	**29** 0355 / 0935 / SU 1617 / 2154	4.3 / 0.1 / 4.2 / 0.2
15 0422 / 1001 / SU 1635 / 2212	3.8 / 0.7 / 3.7 / 0.7	**30** 0439 / 1017 / M 1700 / 2237	4.3 / 0.2 / 4.1 / 0.3
		31 0523 / 1100 / TU 1744 / 2323	4.1 / 0.4 / 3.9 / 0.5

APRIL

Time	m	Time	m
1 0609 / 1145 / W 1830	3.8 / 0.7 / 3.7	**16** 0533 / 1105 / TH 1748 / 2329	3.3 / 1.0 / 3.4 / 1.0
2 0013 / 0658 / TH 1235 / 1922	0.8 / 3.5 / 1.0 / 3.4	**17** 0611 / 1143 / F 1829	3.2 / 1.1 / 3.2
3 0112 / 0757 / F 1337 / 2027	1.1 / 3.2 / 1.3 / 3.2	**18** 0014 / 0657 / SA 1231 / 1920	1.1 / 3.0 / 1.3 / 3.1
4 0229 / 0913 / SA 1458 / 2150	1.3 / 2.9 / 1.5 / 3.0	**19** 0112 / 0758 / SU 1338 / 2028	1.3 / 2.9 / 1.4 / 3.0
5 0359 / 1041 / SU 1626 / 2313	1.4 / 2.9 / 1.5 / 3.1	**20** 0230 / 0918 / M 1505 / 2150	1.3 / 2.9 / 1.5 / 3.0
6 0518 / 1154 / M 1736	1.3 / 3.0 / 1.4	**21** 0355 / 1044 / TU 1629 / 2308	1.2 / 3.0 / 1.3 / 3.2
7 0017 / 0615 / TU 1247 / 1827	3.2 / 1.2 / 3.1 / 1.2	**22** 0508 / 1153 / W 1735	1.0 / 3.2 / 1.0
8 0106 / 0657 / W 1329 / 1907	3.3 / 1.0 / 3.3 / 1.0	**23** 0013 / 0606 / TH 1250 / 1829	3.5 / 0.7 / 3.5 / 0.8
9 0146 / 0733 / TH 1405 / 1943	3.5 / 0.9 / 3.4 / 0.9	**24** 0110 / 0658 / F 1340 / 1919	3.7 / 0.5 / 3.8 / 0.5
10 0221 / 0805 / F 1438 / 2015	3.6 / 0.8 / 3.5 / 0.7	**25** 0201 / 0745 / SA 1427 / 2006	4.0 / 0.3 / 4.0 / 0.3
11 0254 / 0835 / SA 1508 / ○ 2047	3.6 / 0.7 / 3.6 / 0.7	**26** 0249 / 0830 / SU 1513 / ● 2051	4.1 / 0.2 / 4.1 / 0.2
12 0325 / 0905 / SU 1539 / 2117	3.7 / 0.7 / 3.7 / 0.6	**27** 0335 / 0914 / M 1557 / 2136	4.2 / 0.1 / 4.1 / 0.2
13 0356 / 0933 / M 1609 / 2147	3.7 / 0.7 / 3.7 / 0.7	**28** 0420 / 0957 / TU 1640 / 2221	4.1 / 0.3 / 4.0 / 0.3
14 0427 / 1002 / TU 1640 / 2218	3.6 / 0.7 / 3.6 / 0.7	**29** 0505 / 1040 / W 1724 / 2307	3.9 / 0.5 / 3.9 / 0.5
15 0459 / 1032 / W 1713 / 2252	3.5 / 0.8 / 3.5 / 0.8	**30** 0551 / 1124 / TH 1811 / 2357	3.7 / 0.7 / 3.7 / 0.8

PORTUGAL – LISBON
LAT 38°42′N LONG 9°08′W

TIMES AND HEIGHTS OF HIGH AND LOW WATERS

YEAR **1998**

TIME ZONE (UT)
For Summer Time add ONE hour in non-shaded areas

MAY

Day	Time	m	Day	Time	m
1 F	0640 / 1213 / 1901	3.4 / 1.0 / 3.4	16 SA	0559 / 1128 / 1818	3.2 / 1.0 / 3.3
2 SA	0053 / 0735 / 1311 / 2000	1.0 / 3.1 / 1.3 / 3.2	17 SU	0002 / 0647 / 1218 / 1909	0.9 / 3.1 / 1.1 / 3.2
3 SU	0202 / 0842 / 1424 / 2111	1.3 / 2.9 / 1.5 / 3.0	18 M	0058 / 0745 / 1320 / 2011	1.0 / 3.0 / 1.3 / 3.1
4 M	0320 / 1001 / 1545 / 2228	1.4 / 2.8 / 1.5 / 3.0	19 TU	0206 / 0856 / 1437 / 2122	1.1 / 3.0 / 1.3 / 3.2
5 TU	0434 / 1113 / 1655 / 2335	1.3 / 2.9 / 1.4 / 3.1	20 W	0322 / 1012 / 1556 / 2236	1.1 / 3.1 / 1.2 / 3.3
6 W	0533 / 1209 / 1750	1.2 / 3.0 / 1.3	21 TH	0434 / 1122 / 1704 / 2343	0.9 / 3.3 / 1.0 / 3.4
7 TH	0027 / 0619 / 1253 / 1834	3.2 / 1.1 / 3.2 / 1.1	22 F	0536 / 1221 / 1803	0.8 / 3.5 / 0.8
8 F	0109 / 0658 / 1331 / 1912	3.3 / 1.0 / 3.3 / 0.9	23 SA	0043 / 0631 / 1315 / 1856	3.6 / 0.6 / 3.7 / 0.6
9 SA	0147 / 0732 / 1405 / 1947	3.4 / 0.8 / 3.5 / 0.8	24 SU	0137 / 0721 / 1405 / 1946	3.8 / 0.4 / 3.9 / 0.4
10 SU	0223 / 0805 / 1439 / 2020	3.5 / 0.8 / 3.5 / 0.7	25 M	0228 / 0809 / 1452 / 2034	3.9 / 0.3 / 4.0 / 0.3
11 M	0257 / 0836 / 1512 / 2053	3.5 / 0.7 / 3.6 / 0.7	26 TU	0316 / 0854 / 1538 / 2121	3.9 / 0.3 / 4.0 / 0.3
12 TU	0330 / 0907 / 1545 / 2126	3.5 / 0.7 / 3.6 / 0.7	27 W	0403 / 0938 / 1623 / 2206	3.9 / 0.4 / 3.9 / 0.4
13 W	0405 / 0939 / 1620 / 2200	3.5 / 0.7 / 3.6 / 0.7	28 TH	0448 / 1021 / 1707 / 2252	3.7 / 0.5 / 3.8 / 0.5
14 TH	0440 / 1011 / 1656 / 2235	3.4 / 0.8 / 3.5 / 0.7	29 F	0533 / 1105 / 1752 / 2339	3.5 / 0.7 / 3.6 / 0.7
15 F	0518 / 1047 / 1734 / 2315	3.3 / 0.9 / 3.5	30 SA	0619 / 1150 / 1839	3.3 / 0.9 / 3.4
			31 SU	0029 / 0707 / 1241 / 1929	0.9 / 3.1 / 1.2 / 3.2

JUNE

Day	Time	m	Day	Time	m
1 M	0125 / 0802 / 1342 / 2025	1.2 / 2.9 / 1.4 / 3.1	16 TU	0043 / 0730 / 1304 / 1954	0.8 / 3.2 / 1.1 / 3.3
2 TU	0229 / 0905 / 1452 / 2130	1.3 / 2.8 / 1.5 / 3.0	17 W	0143 / 0832 / 1411 / 2057	0.9 / 3.2 / 1.2 / 3.3
3 W	0337 / 1014 / 1603 / 2237	1.4 / 2.8 / 1.4 / 3.0	18 TH	0251 / 0941 / 1524 / 2207	1.0 / 3.2 / 1.2 / 3.3
4 TH	0441 / 1116 / 1704 / 2337	1.3 / 2.9 / 1.4 / 3.0	19 F	0402 / 1051 / 1636 / 2316	1.0 / 3.3 / 1.1 / 3.4
5 F	0534 / 1208 / 1755	1.2 / 3.1 / 1.2	20 SA	0508 / 1155 / 1740	0.9 / 3.4 / 0.9
6 SA	0027 / 0619 / 1252 / 1839	3.1 / 1.1 / 3.2 / 1.1	21 SU	0020 / 0608 / 1253 / 1838	3.5 / 0.7 / 3.6 / 0.7
7 SU	0110 / 0658 / 1331 / 1918	3.2 / 1.0 / 3.3 / 0.9	22 M	0118 / 0702 / 1346 / 1932	3.6 / 0.6 / 3.7 / 0.6
8 M	0151 / 0734 / 1409 / 1955	3.3 / 0.9 / 3.4 / 0.8	23 TU	0212 / 0752 / 1437 / 2021	3.7 / 0.5 / 3.8 / 0.4
9 TU	0230 / 0809 / 1447 / 2031	3.3 / 0.8 / 3.5 / 0.7	24 W	0302 / 0838 / 1524 / 2108	3.7 / 0.5 / 3.9 / 0.4
10 W	0308 / 0844 / 1525 / 2107	3.4 / 0.7 / 3.6 / 0.6	25 TH	0348 / 0922 / 1608 / 2152	3.7 / 0.5 / 3.8 / 0.4
11 TH	0346 / 0919 / 1603 / 2144	3.4 / 0.7 / 3.6 / 0.6	26 F	0432 / 1003 / 1651 / 2234	3.6 / 0.6 / 3.8 / 0.5
12 F	0425 / 0956 / 1642 / 2223	3.4 / 0.7 / 3.6 / 0.6	27 SA	0513 / 1043 / 1731 / 2316	3.5 / 0.7 / 3.7 / 0.7
13 SA	0506 / 1035 / 1724 / 2304	3.4 / 0.8 / 3.6 / 0.6	28 SU	0553 / 1124 / 1812 / 2357	3.3 / 0.9 / 3.5 / 0.8
14 SU	0549 / 1115 / 1809 / 2350	3.3 / 0.9 / 3.5 / 0.7	29 M	0633 / 1207 / 1852	3.2 / 1.0 / 3.3
15 M	0637 / 1200 / 1858	3.3 / 1.0 / 3.4	30 TU	0042 / 0716 / 1256 / 1937	1.1 / 3.0 / 1.2 / 3.1

JULY

Day	Time	m	Day	Time	m
1 W	0133 / 0805 / 1354 / 2029	1.2 / 2.9 / 1.4 / 3.0	16 TH	0119 / 0807 / 1345 / 2033	0.8 / 3.3 / 1.0 / 3.4
2 TH	0233 / 0905 / 1456 / 2131	1.4 / 2.8 / 1.5 / 2.9	17 F	0222 / 0912 / 1456 / 2141	1.0 / 3.2 / 1.1 / 3.3
3 F	0340 / 1012 / 1611 / 2238	1.4 / 2.9 / 1.5 / 2.9	18 SA	0333 / 1023 / 1613 / 2254	1.1 / 3.2 / 1.1 / 3.3
4 SA	0443 / 1115 / 1713 / 2339	1.4 / 2.9 / 1.4 / 2.9	19 SU	0446 / 1133 / 1724	1.0 / 3.3 / 1.0
5 SU	0538 / 1209 / 1804	1.3 / 3.1 / 1.2	20 M	0004 / 0551 / 1237 / 1828	3.3 / 0.9 / 3.5 / 0.9
6 M	0033 / 0624 / 1257 / 1849	3.0 / 1.1 / 3.2 / 1.1	21 TU	0106 / 0648 / 1334 / 1923	3.4 / 0.8 / 3.6 / 0.8
7 TU	0120 / 0706 / 1341 / 1931	3.1 / 1.0 / 3.3 / 0.9	22 W	0201 / 0739 / 1424 / 2012	3.5 / 0.7 / 3.7 / 0.6
8 W	0205 / 0745 / 1424 / 2010	3.2 / 0.9 / 3.5 / 0.7	23 TH	0250 / 0824 / 1510 / 2055	3.6 / 0.6 / 3.8 / 0.5
9 TH	0247 / 0823 / 1506 / 2050	3.4 / 0.8 / 3.6 / 0.6	24 F	0333 / 0905 / 1552 / 2135	3.6 / 0.6 / 3.8 / 0.5
10 F	0329 / 0902 / 1547 / 2129	3.5 / 0.7 / 3.7 / 0.5	25 SA	0412 / 0943 / 1630 / 2212	3.5 / 0.6 / 3.8 / 0.5
11 SA	0411 / 0941 / 1629 / 2209	3.5 / 0.6 / 3.8 / 0.4	26 SU	0449 / 1019 / 1706 / 2247	3.5 / 0.6 / 3.7 / 0.6
12 SU	0453 / 1022 / 1712 / 2251	3.6 / 0.6 / 3.8 / 0.4	27 M	0523 / 1055 / 1740 / 2322	3.4 / 0.7 / 3.6 / 0.8
13 M	0536 / 1105 / 1756 / 2335	3.6 / 0.6 / 3.8 / 0.5	28 TU	0557 / 1131 / 1815 / 2359	3.3 / 0.9 / 3.4 / 1.0
14 TU	0622 / 1151 / 1843	3.5 / 0.8 / 3.7	29 W	0632 / 1211 / 1851	3.2 / 1.1 / 3.2
15 W	0024 / 0711 / 1244 / 1934	0.7 / 3.4 / 0.9 / 3.5	30 TH	0039 / 0712 / 1257 / 1934	1.2 / 3.0 / 1.3 / 3.1
			31 F	0128 / 0801 / 1355 / 2027	1.4 / 2.9 / 1.5 / 2.9

AUGUST

Day	Time	m	Day	Time	m
1 SA	0231 / 0904 / 1509 / 2136	1.5 / 2.8 / 1.5 / 2.8	16 SU	0309 / 1001 / 1557 / 2239	1.2 / 3.2 / 1.2 / 3.1
2 SU	0346 / 1018 / 1626 / 2250	1.5 / 2.8 / 1.5 / 2.8	17 M	0430 / 1118 / 1717 / 2355	1.2 / 3.2 / 1.2 / 3.2
3 M	0455 / 1126 / 1729 / 2357	1.4 / 3.0 / 1.4 / 2.9	18 TU	0541 / 1226 / 1823	1.1 / 3.4 / 1.0
4 TU	0551 / 1223 / 1821	1.3 / 3.1 / 1.2	19 W	0058 / 0639 / 1323 / 1915	3.3 / 1.0 / 3.5 / 0.8
5 W	0052 / 0638 / 1314 / 1907	3.1 / 1.1 / 3.3 / 0.9	20 TH	0150 / 0727 / 1411 / 1959	3.4 / 0.8 / 3.7 / 0.7
6 TH	0142 / 0722 / 1401 / 1950	3.3 / 0.9 / 3.5 / 0.7	21 F	0234 / 0808 / 1453 / 2038	3.5 / 0.7 / 3.8 / 0.6
7 F	0227 / 0803 / 1446 / 2031	3.4 / 0.7 / 3.7 / 0.5	22 SA	0313 / 0846 / 1531 / 2113	3.6 / 0.6 / 3.8 / 0.6
8 SA	0310 / 0844 / 1529 / 2111	3.6 / 0.6 / 3.9 / 0.3	23 SU	0348 / 0920 / 1605 / 2145	3.6 / 0.6 / 3.8 / 0.6
9 SU	0353 / 0925 / 1612 / 2152	3.7 / 0.4 / 4.0 / 0.3	24 M	0420 / 0953 / 1637 / 2217	3.6 / 0.6 / 3.7 / 0.6
10 M	0435 / 1006 / 1654 / 2233	3.8 / 0.4 / 4.0 / 0.4	25 TU	0451 / 1025 / 1708 / 2247	3.5 / 0.7 / 3.6 / 0.8
11 TU	0517 / 1048 / 1738 / 2316	3.8 / 0.4 / 4.0 / 0.4	26 W	0521 / 1057 / 1739 / 2319	3.5 / 0.8 / 3.5 / 0.9
12 W	0601 / 1133 / 1823	3.7 / 0.6 / 3.8	27 TH	0553 / 1131 / 1811 / 2353	3.3 / 1.0 / 3.3 / 1.1
13 TH	0002 / 0649 / 1223 / 1913	0.6 / 3.6 / 0.8 / 3.6	28 F	0628 / 1210 / 1849	3.2 / 1.2 / 3.1
14 F	0054 / 0742 / 1322 / 2010	0.8 / 3.4 / 1.0 / 3.4	29 SA	0033 / 0711 / 1258 / 1936	1.3 / 3.0 / 1.4 / 2.8
15 SA	0155 / 0845 / 1433 / 2119	1.0 / 3.3 / 1.2 / 3.2	30 SU	0125 / 0807 / 1404 / 2040	1.5 / 2.9 / 1.6 / 2.8
			31 M	0241 / 0921 / 1533 / 2204	1.6 / 2.8 / 1.6 / 2.8

PORTUGAL – LISBON

LAT 38°42′N   LONG 9°08′W

TIMES AND HEIGHTS OF HIGH AND LOW WATERS

YEAR 1998

TIME ZONE (UT)
For Summer Time add ONE hour in non-shaded areas

SEPTEMBER

Time	m	Time	m
1 0409 1042 TU 1652 2323	1.6 2.9 1.5 2.9	**16** 0531 1213 W 1813	1.3 3.4 1.1
2 0518 1150 W 1752	1.4 3.1 1.2	**17** 0046 0626 TH 1307 1901	3.3 1.1 3.5 0.9
3 0026 0611 TH 1247 1841	3.1 1.2 3.3 0.9	**18** 0133 0710 F 1351 1940	3.4 0.9 3.7 0.8
4 0118 0658 F 1337 1926	3.3 0.9 3.6 0.6	**19** 0213 0748 SA 1430 2015	3.5 0.8 3.8 0.7
5 0204 0741 SA 1423 2008	3.6 0.6 3.8 0.4	**20** 0248 0823 SU 1505 ● 2047	3.6 0.7 3.8 0.6
6 0248 0823 SU 1508 ○ 2050	3.8 0.4 4.0 0.2	**21** 0321 0855 M 1537 2117	3.7 0.6 3.8 0.6
7 0331 0905 M 1551 2131	4.0 0.3 4.2 0.2	**22** 0351 0926 TU 1607 2146	3.7 0.6 3.7 0.7
8 0413 0947 TU 1634 2212	4.0 0.3 4.2 0.2	**23** 0420 0956 W 1637 2215	3.7 0.7 3.7 0.8
9 0456 1029 W 1718 2255	4.0 0.3 4.1 0.3	**24** 0449 1027 TH 1707 2244	3.6 0.7 3.5 0.9
10 0540 1114 TH 1803 2340	3.9 0.5 3.9 0.6	**25** 0520 1059 F 1739 2316	3.5 1.0 3.4 1.1
11 0626 1204 F 1853	3.7 0.7 3.7	**26** 0554 1135 SA 1815 2351	3.3 1.1 3.3 1.3
12 0030 0719 SA 1302 1950	0.9 3.5 1.0 3.4	**27** 0634 1218 SU 1859	3.2 1.4 3.0
13 0131 0822 SU 1416 2102	1.2 3.3 1.3 3.1	**28** 0037 0725 M 1316 1959	1.5 3.0 1.5 2.8
14 0249 0941 M 1546 2228	1.4 3.2 1.3 3.0	**29** 0144 0834 TU 1440 2123	1.6 2.9 1.6 2.8
15 0417 1104 TU 1710 2346	1.4 3.2 1.3 3.1	**30** 0319 0959 W 1611 2250	1.7 3.0 1.5 2.9

OCTOBER

Time	m	Time	m
1 0442 1116 TH 1719 2357	1.5 3.1 1.2 3.2	**16** 0023 0604 F 1242 1836	3.3 1.2 3.5 1.0
2 0542 1217 F 1812	1.2 3.4 0.9	**17** 0108 0647 SA 1325 1914	3.4 1.1 3.6 0.9
3 0051 0632 SA 1309 1859	3.4 0.9 3.7 0.6	**18** 0146 0724 SU 1403 1947	3.5 0.9 3.7 0.8
4 0139 0717 SU 1358 1943	3.7 0.6 4.0 0.4	**19** 0220 0758 M 1437 2018	3.7 0.8 3.7 0.8
5 0224 0801 M 1444 ○ 2026	4.0 0.3 4.2 0.2	**20** 0252 0830 TU 1509 ● 2048	3.7 0.8 3.7 0.7
6 0308 0844 TU 1529 2108	4.1 0.2 4.3 0.2	**21** 0322 0901 W 1539 2117	3.7 0.7 3.7 0.8
7 0351 0928 W 1613 2151	4.2 0.2 4.3 0.2	**22** 0353 0932 TH 1609 2147	3.7 0.8 3.7 0.8
8 0434 1011 TH 1657 2234	4.2 0.3 4.2 0.4	**23** 0423 1003 F 1641 2216	3.7 0.8 3.5 0.9
9 0519 1057 F 1744 2319	3.9 0.5 3.9 0.6	**24** 0455 1035 SA 1715 2248	3.6 0.9 3.4 1.1
10 0606 1147 SA 1834	3.8 0.7 3.6	**25** 0530 1111 SU 1752 2323	3.5 1.1 3.3 1.3
11 0009 0658 SU 1245 1932	1.0 3.6 1.2 3.3	**26** 0609 1153 M 1836	3.3 1.3 3.1
12 0109 0801 M 1359 2045	1.3 3.3 1.3 3.1	**27** 0008 0658 TU 1247 1933	1.4 3.2 1.4 3.0
13 0227 0919 TU 1528 2210	1.5 3.2 1.4 3.0	**28** 0109 0802 W 1400 2050	1.6 3.1 1.5 2.9
14 0356 1041 W 1650 2326	1.5 3.2 1.3 3.1	**29** 0234 0921 TH 1526 2214	1.6 3.1 1.4 3.0
15 0510 1149 TH 1751	1.4 3.3 1.2	**30** 0401 0617 F 1641 2324	1.5 1.2 3.5 3.2
		31 0509 1144 SA 1740	1.3 3.5 1.0

NOVEMBER

Time	m	Time	m
1 0022 0603 SU 1240 1830	3.5 1.1 3.7 0.7	**16** 0115 0657 M 1332 1918	3.5 1.1 3.5 1.0
2 0112 0652 M 1332 1918	3.8 0.7 4.0 0.4	**17** 0151 0733 TU 1408 1951	3.6 1.0 3.6 0.9
3 0200 0739 TU 1421 2003	4.0 0.5 4.1 0.3	**18** 0224 0807 W 1442 2022	3.7 0.9 3.6 0.9
4 0245 0825 W 1508 ○ 2047	4.2 0.3 4.2 0.3	**19** 0257 0839 TH 1515 ● 2053	3.7 0.8 3.6 0.9
5 0330 0910 TH 1554 2131	4.2 0.3 4.2 0.3	**20** 0330 0912 F 1548 2124	3.7 0.8 3.6 0.9
6 0415 0956 F 1640 2215	4.2 0.3 4.1 0.5	**21** 0403 0944 SA 1622 2155	3.7 0.9 3.5 0.9
7 0501 1043 SA 1727 2300	4.1 0.5 3.9 0.7	**22** 0437 1019 SU 1658 2229	3.7 0.9 3.5 1.0
8 0548 1132 SU 1817 2349	3.9 0.8 3.6 1.0	**23** 0514 1056 M 1737 2306	3.6 1.0 3.4 1.1
9 0639 1228 M 1913	3.7 1.0 3.3	**24** 0555 1138 TU 1821 2351	3.5 1.1 3.2 1.3
10 0045 0738 TU 1334 2018	1.3 3.4 1.3 3.1	**25** 0642 1228 W 1915	3.4 1.2 3.1
11 0155 0846 W 1452 2134	1.5 3.3 1.4 3.0	**26** 0046 0739 TH 1330 2020	1.4 3.3 1.3 3.1
12 0317 1002 TH 1609 2249	1.6 3.2 1.4 3.1	**27** 0158 0847 F 1444 2136	1.5 3.3 1.3 3.1
13 0432 1111 F 1713 2348	1.5 3.3 1.4 3.2	**28** 0319 1001 SA 1559 2248	1.5 3.3 1.2 3.3
14 0531 1207 SA 1802	1.4 3.3 1.2	**29** 0432 1110 SU 1705 2351	1.3 3.5 1.0 3.5
15 0035 0617 SU 1252 1842	3.3 1.2 3.5 1.1	**30** 0534 1212 M 1802	1.1 3.7 0.8

DECEMBER

Time	m	Time	m
1 0046 0629 TU 1308 1854	3.8 0.8 3.9 0.6	**16** 0121 0708 W 1340 1925	3.5 1.2 3.4 1.1
2 0137 0720 W 1401 1943	4.0 0.6 4.0 0.5	**17** 0159 0746 TH 1418 1959	3.6 1.0 3.5 1.0
3 0226 0810 TH 1451 2030	4.1 0.5 4.1 0.4	**18** 0235 0821 F 1454 ● 2033	3.7 0.9 3.5 0.9
4 0314 0857 F 1539 2115	4.2 0.4 4.1 0.4	**19** 0311 0855 SA 1531 2106	3.7 0.9 3.6 0.9
5 0400 0944 SA 1626 2159	4.2 0.4 4.0 0.5	**20** 0347 0930 SU 1608 2140	3.8 0.8 3.6 0.9
6 0446 1030 SU 1712 2243	4.1 0.5 3.8 0.7	**21** 0424 1006 M 1646 2216	3.8 0.8 3.6 0.9
7 0532 1117 M 1758 2329	4.0 0.7 3.6 0.9	**22** 0503 1043 TU 1726 2254	3.8 0.8 3.5 1.0
8 0619 1206 TU 1847	3.8 1.0 3.4	**23** 0544 1124 W 1809 2337	3.7 0.9 3.5 1.1
9 0018 0709 W 1300 1940	1.2 3.5 1.2 3.2	**24** 0629 1211 TH 1857	3.6 1.0 3.4
10 0115 0804 TH 1402 2041	1.4 3.3 1.4 3.0	**25** 0027 0719 F 1304 1953	1.2 3.5 1.1 3.3
11 0223 0907 F 1511 2150	1.6 3.2 1.5 3.0	**26** 0128 0818 SA 1408 2059	1.3 3.4 1.2 3.3
12 0337 1016 SA 1620 2258	1.6 3.1 1.5 3.1	**27** 0239 0926 SU 1520 2211	1.4 3.4 1.2 3.3
13 0445 1120 SU 1718 2354	1.5 3.2 1.4 3.2	**28** 0356 1038 M 1632 2321	1.3 3.4 1.1 3.5
14 0541 1213 M 1807	1.4 3.3 1.3	**29** 0507 1147 TU 1738	1.2 3.5 1.0
15 0040 0628 TU 1259 1848	3.3 1.3 3.3 1.2	**30** 0023 0610 W 1250 1835	3.6 1.0 3.7 0.8
		31 0120 0707 TH 1347 1928	3.8 0.8 3.8 0.7

SINES IG 41

Baijo Alentejo 37°56'·95N 08°52'·06W (marina ent)
CHARTS
AC 3276, 3636; PC 84, 39; SC 4311, 43A; SHOM 6995
TIDES
Standard Port LISBOA (←); ML 2·0; Zone 0 (UT)

Times				Height (metres)			
High Water		Low Water		MHWS	MHWN	MLWN	MLWS
0400	0900	0400	0900	3·8	3·0	1·4	0·5
1600	2100	1600	2100				
Differences SINES							
−0050	−0030	−0020	−0010	−0·4	−0·4	0·0	+0·1
MILFONTES (37°43'·0N 08°47'·0W)							
−0040	−0030	No data		−0·1	−0·1	+0·1	+0·2
ARRIFANA (37°17'·5N 08°52'·0W)							
−0030	−0020	No data		−0·1	0·0	0·0	+0·2

SHELTER
Good in small marina protected by new mole at E end of the bight Vasco da Gama (FVs are at W end); or ⚓ NW of marina in 4m. Nearby Old town is pleasant, away from commercial terminals in NW and SE parts of the large hbr. Sines is a useful haven, strategically located 51M SSE of Cascais and Lisbon and 57M N of Cape St Vincent.
NAVIGATION
WPT 37°56'·03N 08°53'·18W, PHM buoy (Fl R 3s) 225°/045° from/to ent to ⚓ or marina, 1·25M. From the NW, caution: the southern 800m section of the W mole is derelict and partly covers. The ☆ (Fl 3s 20m 12M) on it is 400m N of the unlit mole end; the WPT buoy is 270m S of the mole end and <u>must</u> be rounded; no short cuts! In strong S'lies swell may enter, causing turbulence off the W mole.
LIGHTS AND MARKS
Cabo de Sines lt ho, Fl (2) 15s 50m 26M, is 8ca NW of the marina; it is obsc'd by oil tanks 001°-003° and 004°-007°. Only ldg lts are for commercial terminals. The elbow (off chartlet) of the E Mole has ☆ L Fl Y 8s 16m 9M. From afar, an oil refinery flare in the town and several tall chy's/masts 3-4M to the E (all with R lts) help to locate the port.
RADIO TELEPHONE
Call *Porto de Sines* (Port Authority) VHF Ch 11, 13, 16.
TELEPHONE (Dial code 069)
Hr Mr 860600, 📠 860690.
FACILITIES
Marina (80, inc some Ⓥ); contact via Hr Mr; AC, FW; (planned to expand); **Club Náutico** close SE of marina; **Services:** D & P from FV dock near HW; ME, EI.
Town: R, V, Bar, Gaz, Ⓑ, ✉, ⇌, ✈ Lisboa or Faro.

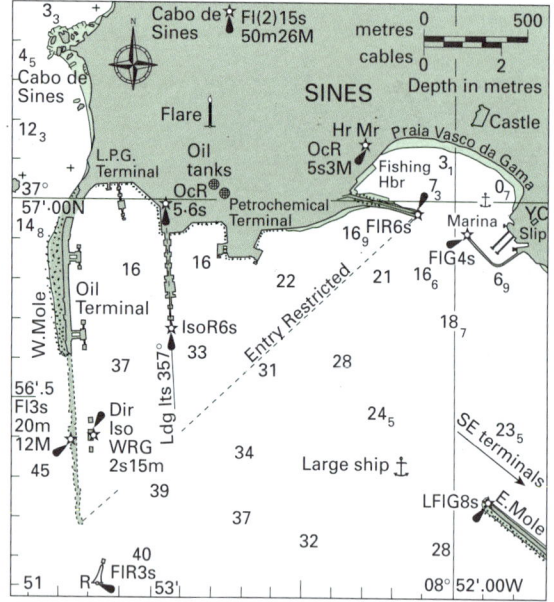

LAGOS IG 42

Algarve 37°05'·84N 08°39'·90W
CHARTS
AC 3636, 89; PC 88, 40, 41, 7, 24; SC 43A, 44A; SHOM 3388
TIDES
Standard Port LISBOA (←); ML 2·0; Zone 0 (UT)

Times				Height (metres)			
High Water		Low Water		MHWS	MHWN	MLWN	MLWS
0400	0900	0400	0900	3·8	3·0	1·4	0·5
1600	2100	1600	2100				
Differences LAGOS							
−0100	−0040	−0030	−0025	−0·4	−0·4	0·0	+0·1
ENSEADA DE BELIXE (Cape St. Vincent)							
−0050	−0030	−0020	−0015	+0·3	+0·2	+0·3	+0·3

SHELTER
Very good in marina (dredged 3m) on E bank, 7ca up-river from hbr ent, access H24 all tides. 3kn speed limit in hbr. The ⚓ close NE of the E Mole in 3-5m is exposed only to E and S winds and possible SW swell.
NAVIGATION
WPT 37°05'·76N 08°39'·58W, 104°/284° from/to hbr ent, 2½ca. From Ponta da Piedade, Fl 7s 49m 20M, 1·2M to the S, keep 5ca offshore to clear rocks; otherwise there are no offshore dangers. Final appr is on 284°, W mole head lt, Fl R 6s, in transit with Santo António church. Berth at arrival pontoon stbd side just before lifting foot-bridge which opens on request 0800-2200 June to mid-Sept (0900-1900 winter), except for some ¼hr spells for train passengers. R/G tfc lts, but unmasted craft may transit when bridge is down.
LIGHTS AND MARKS
Lts as chartlet. The river chan is unlit, apart from shore lts. Alvor lt, L Fl R 6s 31m 7M, is 3·5M to ENE; Ponta do Altar (Portimão), L Fl R 5s 30m 14M, is 7M E.
RADIO TELEPHONE
Call *Marina de Lagos* VHF Ch 62 for bridge opening.
TELEPHONE (Dial code 082)
Hr Mr, 📠, Met: as marina; Ⓗ 763034; Police 762930; Fire 760115; Brit Consul 417800.
FACILITIES
Marina (462+ Ⓥ, max LOA 20m), 3420 escs, ☎ 762765, 📠 761588, P, D, AC, FW, Slip, BH (30 ton), BY, ME, EI, Sh, V, R, Bar, CH, Ⓑ, ✉; **Club de Vela** ☎ 762256; **Services:** usual amenities, ⇌, ✈ Faro (75 km) ☎ (089) 818281.

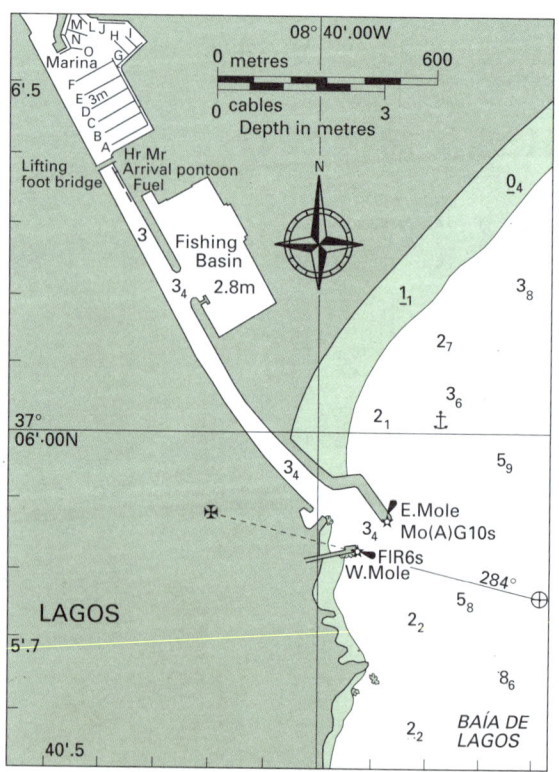

PORTIMÃO
IG 43

Algarve 37°06'·60N 08°31'·64W

CHARTS
AC 83, 3636, 89, 92; PC 41, 7, 24; SC 43A, 44A; SHOM 3388

TIDES
Standard Port LISBOA (←); ML 2·0; Zone 0 (UT)

TIDES continued

Times				Height (metres)			
High Water		Low Water		MHWS	MHWN	MLWN	MLWS
0400	0900	0400	0900	3·8	3·0	1·4	0·5
1600	2100	1600	2100				
Differences PORTIMÃO							
–0100	–0040	–0030	–0025	–0·5	–0·4	0·0	+0·2
PONTA DO ALTAR (Lt ho E of ent)							
–0100	–0040	–0030	–0025	–0·3	–0·3	0·0	+0·1

SHELTER
Good on pontoons off W bank, 2M up-river from hbr ent, just before 2 low road/rail bridges. Caution: flood & ebb streams run hard. Good ⚓ inside E mole in about 4m, but prone to wash from FVs and swell possible. Note: a new marina is planned on the W bank, as shown; date unknown. by mid-1999

NAVIGATION
WPT 37°06'·40N 08°31'·73W, 201°/021° from/to hbr ent, 2ca. There are no offshore hazards; entrance and river are straightforward.

LIGHTS AND MARKS
Pta do Altar lt ho is on low cliffs to E of ent. The R-roofed church at Ferragudo is conspic by day, almost on ldg line 021°. Ldg lts: front Oc R 5s 17m 8M; rear, Oc R 7s 33m 8M; both on R/W banded columns, moved to suit chan.

RADIO TELEPHONE
Port VHF Ch 11 16.

TELEPHONE (Dial code 082)
Hr Mr 417714; Pontoon, see below; # 24239; Police 22440; Consul 417800.

FACILITIES
Pontoon (about 60+ few Ⓥ) ☎ 23074, 417255, AC, FW; Services: P & D (cans), CH, El, ME, BY, Gaz, R, V, Bar, Ⓑ, ✉; ⛴, ✈ Faro (65 km).

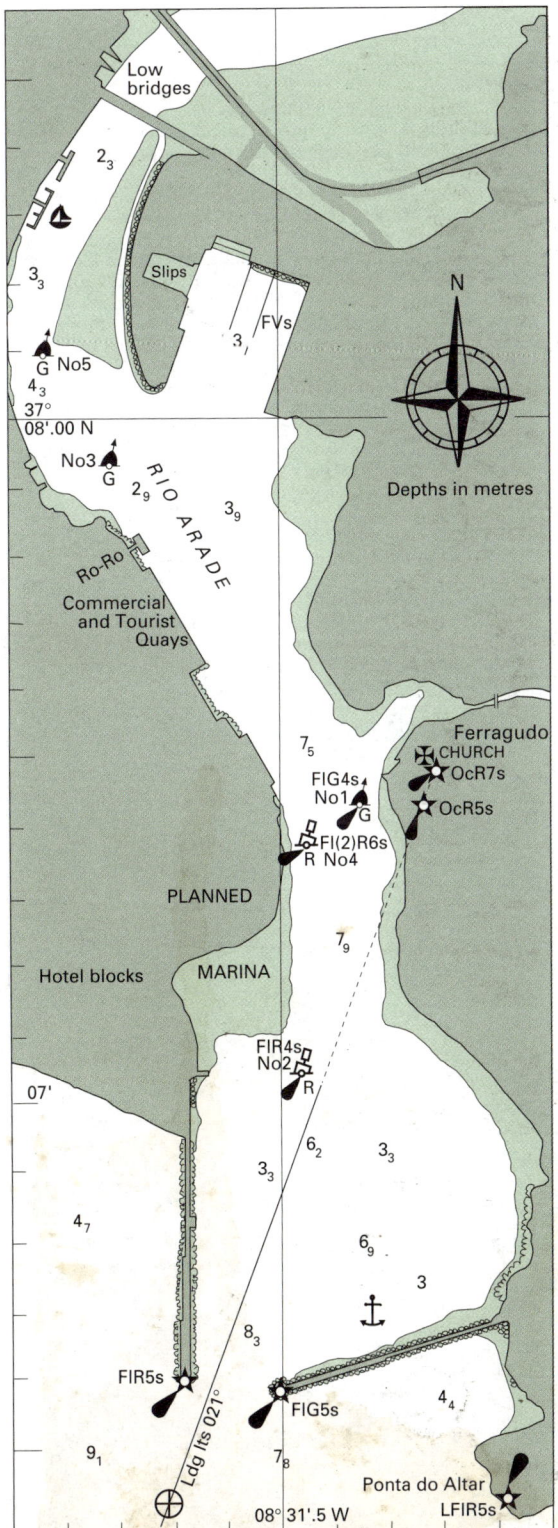

VILAMOURA
IG 44

Algarve 37°04'·10N 08°07'·35W

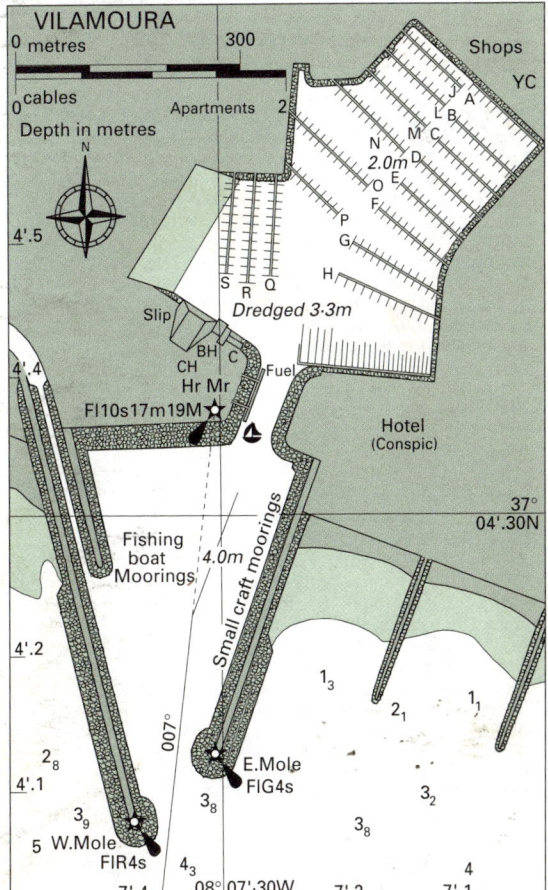

VILAMOURA continued
CHARTS
AC 89, 92; PC 8, 24, 42; SC 44A; SHOM 3388
TIDES
Standard Port LISBOA (←—); ML 2·0; Zone 0 (UT)

Times				Height (metres)			
High Water		Low Water		MHWS	MHWN	MLWN	MLWS
0400	0900	0400	0900	3·8	3·0	1·4	0·5
1600	2100	1600	2100				

Differences ENSEADA DE ALBUFEIRA (6M to the W)
−0035 +0015 −0005 0000 −0·2 −0·2 +0·1 +0·2
CABO DE SANTA MARIA (Faro/Olhão, 14M to the ESE)
−0050 −0015 +0005 0000 −0·4 −0·4 0·0 +0·1
VILA REAL DE SANTO ANTÓNIO (Portugal/Spain border)
−0050 −0015 −0010 0000 −0·4 −0·3 0·0 +0·2

SHELTER
Very good in marina dredged 2·0 – 3·3m, max LOA 40m. In strong S'lies the 100m wide ent can be dangerous. On arrival berth at reception pontoon (⚓); for a night stop, it is possible to check in/out at the same time. Procedures are said to be more streamlined than of old.
NAVIGATION
WPT 37°03'·60N 08°07'·40W, 187°/007° from/to hbr ent, 5ca. There are groynes to the E, but no offshore dangers.
LIGHTS AND MARKS
Marina is surrounded by apartment bldgs; large hotel is conspic on E side of ent. Main lt, Fl 10s 17m 19M, lattice tr with R top on Hr Mr's bldg; this lt between the W and E mole head lts leads 007° into the outer hbr.
RADIO TELEPHONE
Call marina *Vilamoura Radio* VHF Ch 62 (0900-2100/1800 off season) 16. Weather forecast for Algarve coast is broadcast in English and Portuguese on Ch 20 at 1000; a bulletin for all Portugal is posted daily.
TELEPHONE (Dial code 089)
Hr Mr, # & Met: as marina, below; LB 313214; ⊞ 802555 (Faro); Police 388989.
FACILITIES
Marina (1000 inc Ⓥ), 2700 escs, ☎ 302923/7, 🛥 302928, AC, FW, P & D, Slip, BH (60 ton), C (2 & 6 ton), ME, EI, Ⓔ, BY, CH, Gaz, Sh, SM; **Club Náutico** Bar, R, ▣;
Faro (20 km) most facilities; Ⓑ, ⇌, ✈.

OTHER HARBOURS AND ANCHORAGES EASTWARD TO THE PORTUGUESE/SPANISH BORDER

FARO and OLHÃO. Ent 37°57'·90N 07°52'·11W. AC 83; PC 91, 92; SC 44A/B. Tide, see above (Cabo de Santa Maria). The narrow ent through sand dunes leads into lagoons amongst salt marshes. Appr on 352° to clear shoals W of ent between moles, Fl R and G 4s respectively. Cabo de Santa Maria lt ho at root of long E mole is conspic, Fl (4) 17s 49m 25M. It is also rear ldg lt 021° once inside ent; front ldg lt, Barra Nova Oc 4s 8m 6M. Abeam C. de Sta Maria, fork WNW to Faro or NE to Olhão. The **Faro** chan carries 5·4m, but depths change; it is buoyed/lit for the 4M to the Commercial quay (unsuitable for AB). Appr near HW to ⚓ in 3m, 3ca SW of Doca de Recreio (low bridge across ent). The chan to **Olhão** (5M) is shallower, 2·4m. ⚓ off Ponte Cais (Ilha da Culatra) or off Olhão yacht hbr, for small craft only. Both towns are sizeable with usual facilities, inc ⇌, ✈.

VILA REAL DE SANTO ANTÓNIO and AYAMONTE.
37°09'·00N 07°23'·35W. AC 89; PC 97; SC 4411. Tide, IG 44. Vila Real is on the W bank (Portugal) of Rio Guadiana and Ayamonte on E bank (Spain). The former had a marina (2·2m) at NW corner of FV basin, 2·5M upriver from ent. NB: New marina (300 berth) at about 37°11'·6N on W bank was to have been opened Dec 1997.
Ayamonte has a yacht pontoon (about 1m) in basin to S. Best entry to river between training walls at HW −3 over bar (2·5m) marked by 2 SHM and 2 PHM lt buoys; lat/long of No 1 SHM buoy, Q (3) G 6s, is given above. High bldgs, FR, are 3M WNW of ent. White suspension bridge 2M N of towns is visible from seaward. Other lts: Vila Real lt ho, Fl 6·5s 51m 26M, W ○ tr, B bands. E trng wall (partly covers) bn, Fl G 3s; W trng wall hd, Fl R 5s. *Porto de Vilareal* VHF Ch 11, 16; 2484kHz. Facilities: Vila Real, FW, D, YC, R, Slip; Ayamonte, D, YC. Both towns: V, R, Bar, Ⓑ, ✉, ✈ (Faro).

ISLA CRISTINA   IG 45
Huelva 37°11'·88N 07°19'·65W
CHARTS
AC 89; SC 4411, 441, 44B; SHOM 7300
TIDES
Standard Port LISBOA (←—); ML 1·8; Zone −0100

Times				Height (metres)			
High Water		Low Water		MHWS	MHWN	MLWN	MLWS
0500	1000	0500	1100	3·8	3·0	1·4	0·5
1700	2200	1700	2300				

Use Differences AYAMONTE (4M WNW)
+0005 +0015 +0025 +0045 −0·7 −0·6 0·0 −0·1

SHELTER
Good once inside marina (2·5–3m); the river ent is open to SE winds (like many hbrs on the Algarve). FVs berth at quays to the N of marina. There is little space to ⚓.
NAVIGATION
WPT 37°10'·75N 07°19'·25W (off chartlet), 133°/313° from/to chan ent lts, 3ca. Chan runs NW for nearly 1M, then curves to stbd, past No 1 SHM lt buoy. Stand on until tall conspic W bldg (looks like a lt ho) bears about 100°, then alter for it. At a small W buoy (uncharted) just short of this bldg, turn NE for No 2 PHM lt buoy, close W of marina. Best appr at half-flood, to see drying sandbanks. Chan beyond ldg lts is prone to change depth/direction.
LIGHTS AND MARKS
Ldg lts (front Q 7m 5M; rear Fl 4s 12m 5M) lead approx 313°, between the W mole head VQ (2) R 5s 7m 4M and bn, Fl G 2s 4m 5M, on head of drying trng bank to stbd. The powerful lt, Fl 6·5s 51m 26M, W tr + B bands, at Vila Real de Santo António is 4M to the W.
RADIO TELEPHONE
Marina VHF Ch 09.
TELEPHONE (Dial code 959)
Hr Mr, #, Met: via marina.
FACILITIES
Marina (202 AB, inc Ⓥ), 1553 ptas, ☎/🛥 343501, FW, AC, D, Slip, BH (32 ton), ME, EI, Sh, R, Bar, Ice, ▣;
Note: marina is first (E-bound) of 7 modern marinas run by Junta de Andalucía. **Town**: V, R, Bar, Ⓑ, ✉; Bus Ayamonte (17 km); ✈ Faro (65 km).

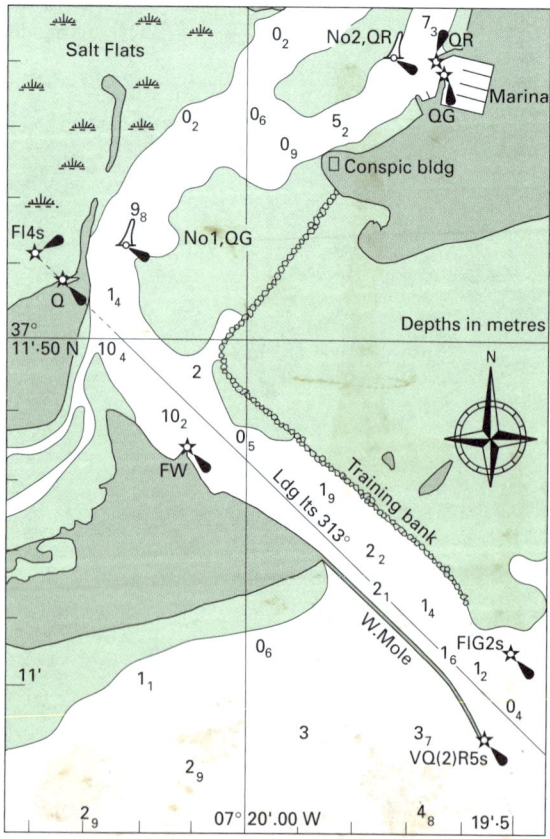

MAZAGÓN IG 46

Huelva 37°08'·00N 06°50'·00W

CHARTS
AC 83, 90, 92; SC 4413, 441, 442; SHOM 6862, 7300
TIDES
Standard Port LISBOA (←); ML 1·8; Zone –0100

Times				Height (metres)			
High Water		Low Water		MHWS	MHWN	MLWN	MLWS
0500	1000	0500	1100	3·8	3·0	1·4	0·5
1700	2200	1700	2300				
Differences RÍA DE HUELVA, BAR							
0000	+0015	+0035	+0030	–0·6	–0·5	–0·2	–0·1

SHELTER
Good in large, modern marina, ent facing WNW up-river. Inner approach is sheltered by training wall on SW side.

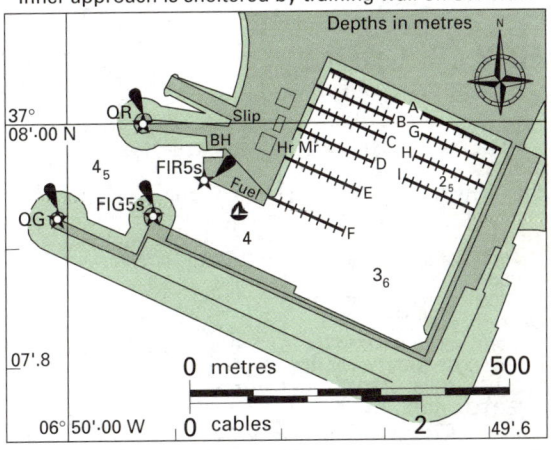

NAVIGATION
WPT 37°05'·65N 06°49'·05W, WCM buoy, Q (9) 15s, 159°/339° from/to marina, 2·5M. From the W, be aware of an oil pipe line running N/S, ending about 5M W of the WPT; it is marked by 4 SPM buoys, each Fl (4) Y 20s. The appr is straightforward via well lit, buoyed, dredged (10m) chan 339°, which bears away NW to Huelva industrial complex. After the first pair of chan buoys, the very long training wall lies to port; its seaward end has lt twr, Fl (3+1) WR 20s 30m 12/9M. Keep clear of merchant ships.
LIGHTS AND MARKS
Ldg lts 339°: front, Q 49m 8M; rear, Fl R 2s 55m 8M, both on R/W chequered trs. After No 7 SHM buoy, Fl (4) G 20s, alter stbd for the marina ent. Picacho lt ho, Fl (2+4) 30s 51m 25M, W tr with R corners, is 5ca NE of marina ent.
RADIO TELEPHONE
Marina VHF Ch 09 16. Huelva Port Ch 06 11 12 14 16.
TELEPHONE (Dial code 959)
Marina, see below; ⌗ & Met via marina.
FACILITIES
Marina (516 AB in 2·5-4m, inc Ⓥ), 1553Ptas, ☎ & ⛽ 376237, FW, P, D, AC, Slip, BH (32 ton), ME, El, Sh, Ice, YC, Bar, R; ⇌ Huelva (24km); ✈ Sevilla (130km).

HARBOURS CLOSE WEST OF RIA DE HUELVA

RIO DE LAS PIEDRAS. SC 4412. SWM buoy, L Fl 10s, at 37°11'·82N 07°02'·35W marks ent; thence steer NW past SHM buoy, Fl G 5s, and PHM buoy, Fl (2) R 6s, into river. Buoys are moved to suit shifting chan (0·6m) and shoals. ⚓ in river is sheltered by long spit. El Rompido, lt ho Fl (2) 10s, is 4M to W with YC, ☎ (959) 399349, ⛽ 399217; BY.

PUNTA UMBRIA. AC 83. SC 4413. SCM buoy, VQ (6)+L Fl 10s, at 37°09'·64N 06°57'·09W marks chan ent (1·1m). NB: the oil pipe line described under IG 46 is close E. Bkwtr head, also VQ (6)+L Fl 10s, is to port. YC, dark R bldg, and pontoon is 1M up-river, Fl (3) G. Note: all 3 lts on SW side are G. Temp'y AB, M or ⚓ if any room. YC ☎ (959) 311899.

CHIPIONA IG 47

Cádiz 36°45'·00N 06°25'·63W

CHARTS
AC 85; SC 4421, 4422 (18 sheets); SHOM 6342, 5365
TIDES
Standard Port LISBOA (←); ML 1·9; Zone –0100

Times				Height (metres)			
High Water		Low Water		MHWS	MHWN	MLWN	MLWS
0500	1000	0500	1100	3·8	3·0	1·4	0·5
1700	2200	1700	2300				
Differences RIO GUADALQUIVIR, BAR							
–0005	+0005	+0020	+0030	–0·6	–0·5	–0·1	–0·1
BONANZA (36°48'N 06°20'W)							
+0025	+0040	+0100	+0120	–0·8	–0·6	–0·3	0·0
CORTA DE LOS JERONIMOS (37°08'N 06°05'W)							
+0210	+0230	+0255	+0345	–1·2	–0·9	–0·4	0·0
SEVILLA							
+0400	+0430	+0510	+0545	–1·7	–1·2	–0·5	0·0

SHELTER
Good in modern marina, but swell intrudes in NW gales. Yachts berth on SE side in 2·5m, larger yachts in first basin on port side. FVs berth against the NW bkwtr.
NAVIGATION
WPT 36°45'·80N 06°26'·83W, No 1 SWM buoy, L Fl 10s, 308°/128° from/to marina ent, 1·33M. Caution: extensive fish havens with obstructions 2·5m high lie from 3·5M to 7·5M NW of the WPT. Same WPT is on ldg line 069° for buoyed chan into Rio Guadalquivir.
LIGHTS AND MARKS
Pta de Perro lt ho is conspic 1M SW of marina. 1·7M W of this lt ho, the drying Bajo Salmedina (off chartlet) which must be rounded if coming from S, is marked by WCM bn tr, Q (9) 15s. Inside hbr bkwtr, ☆ Fl (3) G 9s, near FV berths, is not visible from seaward. Ldg lts 069° for Rio Guadalquivir: front, Q 27m 10M; rear, Iso 4s 60m 10M.
RADIO TELEPHONE
Marina VHF Ch 09. Port and Rio Guadalquivir Ch 12.
TELEPHONE (Dial code 956)
Hr Mr, ⌗ and Met: as marina.

FACILITIES
Marina (358 AB inc Ⓥ), 1553Ptas, ☎ 373844, ⛽ 370037, AC, P, D, FW, Slip, BH (32 ton), CH, ME, El, Sh, Bar, R, ⌕; **Town**: V, R, Bar, Ⓑ, ✉ and ⇌ and ✈ Jerez (32 km).

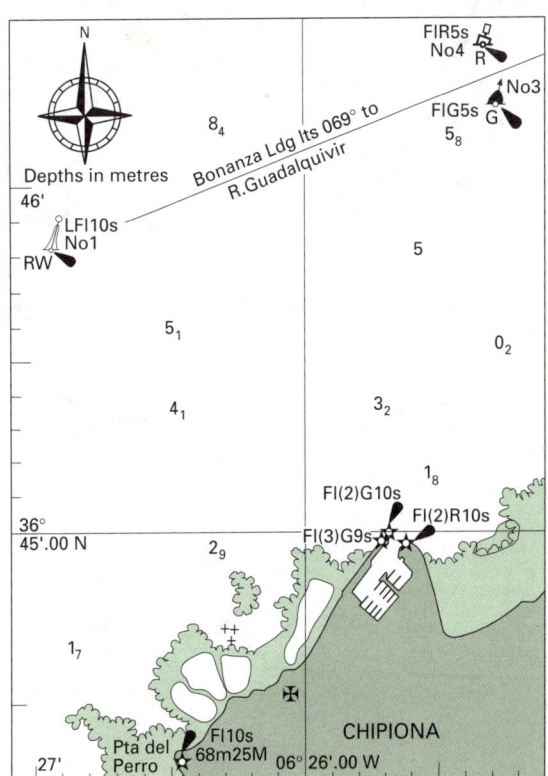

RIO GUADALQUIVIR.

The river is not difficult to navigate but can be hot and uninspiring, through almost featureless terrain. Updated AC 85 or SC 4422 essential as buoys/bns liable to change. The lit, buoyed channel starts 1·4M E of the Fairway buoy which is also WPT for CHIPIONA (IG 47). From here to a lock 2M S of Sevilla is approx 50M. At Bonanza (36°48'N 06°20'W) the channel becomes truly riverine, 750m wide, 250m nearer Sevilla. Depths are rarely less than 5m, best water usually being on the outside of bends. The flood runs for about 7hrs (3kn sp, 1kn nps), the ebb for 5½ hrs, so that it should be possible to make the lock on one tide, passing Bonanza at LW–½hr. There are no recognised stopping places and any ⚓ is vulnerable to passing traffic; monitor VHF Ch 12. The lock, below HT cables (44m) between conspic R/W pylons, gives access to Sevilla (IG48), 2M N.

SEVILLA IG 48

Sevilla 37°20'·00N 05°59'·65W (Lock)

CHARTS & TIDES
See IG47. ML Sevilla 1·3M; Zone –0100

SHELTER
The 3 marinas are: **Gelves**, 1·5M N of Bn 52 on the W arm of the river. It is badly silted and approx 3M from the city. **Marina Yachting Sevilla**, 300m E of the lock almost below HT cables, one long pontoon in quiet creek 3M from city; useful stopover prior to transiting road/rail bridge 2M N. **Club Náutico de Sevilla**, excellent facilities close to city.

NAVIGATION
See Rio Guadalquivir. The lock, 2M S of city centre, opens H24 approx every 3 hrs (check at Chipiona); secure to ladders/rubber strips. After conspic high suspension bridge (ring road), negotiate a road/rail bridge, opening 0800 & 2000 in season (0800 & 1800 winter); check times at lock. Beyond that, bascule bridge is permanently open.

RADIO TELEPHONE
Marinas VHF Ch 09. Port, lock and road/rail bridge Ch 12.

TELEPHONE (Dial code 95; see below. Brit Consul 4228875).

FACILITIES
Gelves marina (150+ Ⓥ), 650pta ☎ 5760728, ✉ 5760464, about 1·5m, waiting pontoon in river, AC, FW, CH, ME, BH (25 ton), ▢, R, V, Bar; **Marina Yachting Sevilla** (400+ Ⓥ), 6m, ☎ 4230326, ✉ 4230172, Slip, AC, FW, ME, ▢; **Club Náutico Sevilla** (100+ Ⓥ, pre-booking advised), ☎ 4454777, ✉ 4284693, AC, FW, D & P, Slip, R, Bar, Ice, ▢.
City: All amenities; ⇌; ✈ (10 km).

BAY OF CÁDIZ IG 49

Cádiz 36°33'·50N 06°18'·70W.

CHARTS
AC 86, 90; SC 443A, 443B, 443; SHOM 6877, 5365

TIDES
Standard Port LISBOA (←); ML 1·8; Zone –0100

Times				Height (metres)			
High Water		Low Water		MHWS	MHWN	MLWN	MLWS
0500	1000	0500	1100	3·8	3·0	1·4	0·5
1700	2200	1700	2300				
Differences ROTA							
–0010	+0010	+0025	+0015	–0·7	–0·6	–0·3	–0·1
PUERTO DE SANTA MARIA							
+0006	+0006	+0027	+0027	–0·6	–0·4	–0·3	–0·1
PUERTO CÁDIZ							
0000	+0020	+0040	+0025	–0·5	–0·5	–0·2	0·0

SHELTER
Good in all 4 marinas: Rota, Puerto Sherry, Puerto de Sta Maria (where very pleasant Real Club Náutico has a few Ⓥ berths), and Puerto America (at N tip of Cádiz).

NAVIGATION
Rota WPT, 36°36'·00N 06°21'·00W, 185°/005° from/to marina ent, 1M. Al Fl WG 9s airfield lt and water tank are almost in transit 005°. From W/NW keep 1·5M offshore to clear shoals. Naval vessels may ⚓ S of Rota Naval base.
Puerto Sherry WPT, 36°34'·30N 06°15'·42W, bears 210°/030° from/to marina ent, 5ca.
For **Puerto de Santa Maria**, same WPT bears 258°/078° from/to W training wall head (Fl R 5s), 5ca (at mouth of canalised Rio Guadalete, dredged 4·5m). From W, the N Chan trends ESE, inshore of shoals toward both marinas.
Puerto America (Cádiz) WPT, 36°33'·50N 06°18'·70W, 297°/117° from/to Dique de San Felipe head, 1·8M. From S, keep to seaward of extensive shoals N & W of Cádiz; appr via the Main Chan. The shore is generally low-lying.

LIGHTS AND MARKS
Lts/buoys as shown; Bn No 3 ✯ is reported inop for last 7 years. Rota lt ho, Oc 4s, W tr + R band, overlooks marina. Conspic daymarks include: Puerto Sherry, W bkwtr hd ◯ twr. Cádiz city, golden-domed cathedral and radio twr (113m) close SE; two cable pylons (154m) in docks area. W bldg on Dique de San Felipe near Puerto America.

RADIO TELEPHONE
Marinas VHF Ch 09. Cádiz port 11, 12, **14**, 16.

TELEPHONE (Dial code 956)
Hr Mr 224011/224005; ⌗ & Met via marinas (see below).

FACILITIES
Rota Marina (362 AB, inc Ⓥ; 2·5-3m), 1553Ptas, ☎ & ✉ 813811, AC, P, D, FW, Slip, BH (32 ton), ME, El, C (5 ton);
Puerto Sherry Marina (753 AB, inc Ⓥ; 4m), ☎ 870103, ✉ 873902, AC, P, D, FW, Slip, BH (50 ton), ME, El, C, SM, Ⓔ, CH, Sh, ▢, R, Bar, YC; Arrivals berth to port inside the ent.
Real Club Náutico de Puerto de Santa Maria (175 AB+few Ⓥ); 10 hammerhead pontoons (A-K) on NW bank plus two mid-stream pontoons. ☎ 852527, ✉ 874400; 1000Ptas, AC, P, D, FW, Sh, Slip, ME, C (5 ton), BH (25 ton), Slip, R, Bar;
Puerto America Marina (270 AB, inc Ⓥ; 8m), 1553Ptas, ☎ & ✉ 224240, AC, P, D, FW, Slip, ME, El, C (10 ton); close SW is **Real Club Náutico de Cádiz** ☎ 253903; R, Bar, D.
City: all amenities, Hydrographic Office, ⇌; ✈ Jerez (25 km).

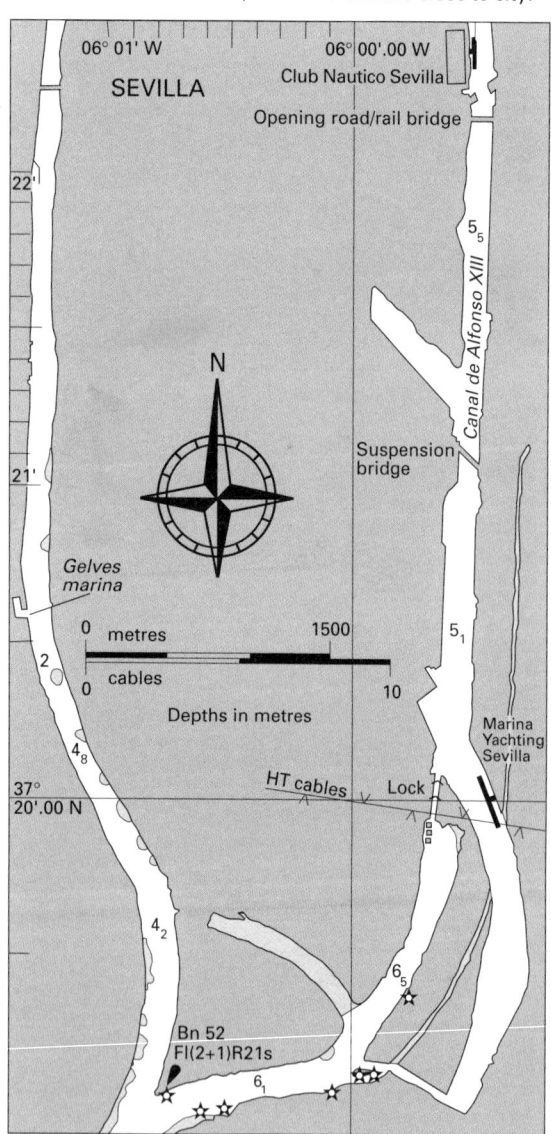

Southwest Spain

BAY OF CÁDIZ including marinas at: ROTA, PUERTO SHERRY, PUERTO DE SANTA MARIA and PUERTO AMERICA

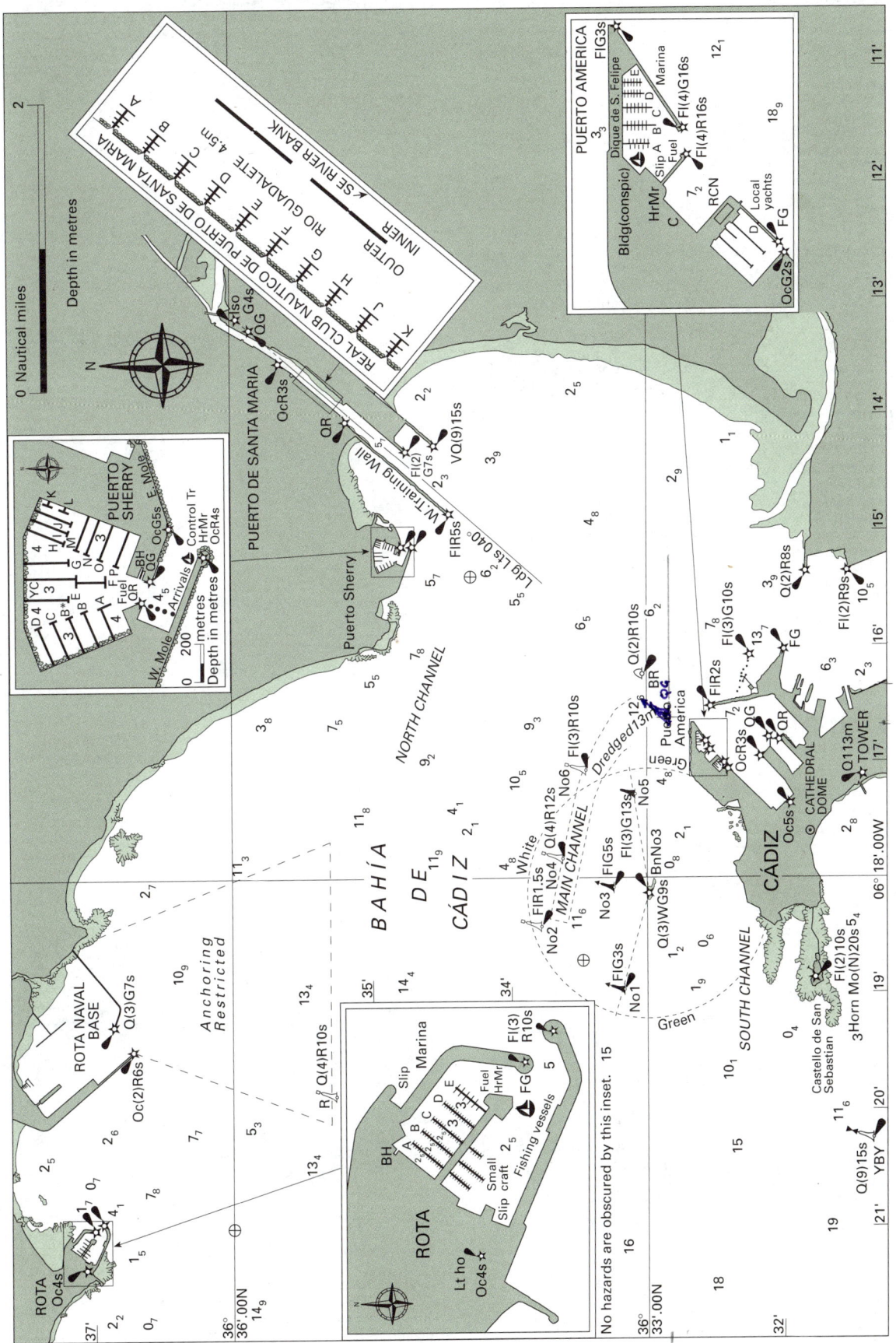

SANCTI-PETRI IG 50

Cádiz 36°23'·80N 06°12'·50W

CHARTS
AC 90; SC 4438, 443; SHOM 5365

TIDES
Interpolate between differences Puerto Cádiz (IG 49) and Cabo Trafalgar (IG 51); ML No data; Zone –0100

SHELTER
Good, except in S'lies. ⚓ or moor in stream, W of marina; AB possible in quoted 3m depth; see Hr Mr. Pontoons to N of YC may be renewed/extended in the future.

NAVIGATION
WPT 36°22'·40N 06°13'·05W, 230°/050° from/to turn point for 2nd set of ldg lts, 5ca. El Arrecife, a long drying reef, and other shoals bar a direct appr from W/NW. Both sets of ldg lts/trs are hard to see by day, but the PHM/SHM lt bns (Fl R/G 5s) are easily seen; careful fixing on these and the castle will help pilotage (night appr not advised). Chan buoys are laid in season. Best appr at half-flood in fair vis; least charted depth 2·2m. Sp tide reaches 2½-3kn. Swell and/or strong S'lies render the appr dangerous.

LIGHTS AND MARKS
The castle with 16m □ tr, Fl 3s, is conspic on islet to W of approach chan. First ldg lts 050° (off chartlet): front, Fl 5s 12m 6M; rear, Oc (2) 6s 16m 6M, both on lattice trs. Second ldg lts 346·5° (Pta del Boquerón): front Fl 5s 11m 6M; rear, Oc (2) 6s 21m 6M, both on lattice trs.

RADIO TELEPHONE
Marina *Puerto Sancti-Petri* VHF Ch 09.

TELEPHONE (Dial code 956)
Hr Mr 495434.

FACILITIES
Marina (approx 200, inc ⓥ), ☎ 495434, AC, FW, Slip, C, limited CH and V in season. Note: marina is under development by Junta de Andalucía. **Club Náutico** Bar, R.
Village: No facilities in abandoned "ghost town" at mouth of sandy, peaceful lagoon; ≷ San Fernando (18 km); ✈ Jerez (50 km).

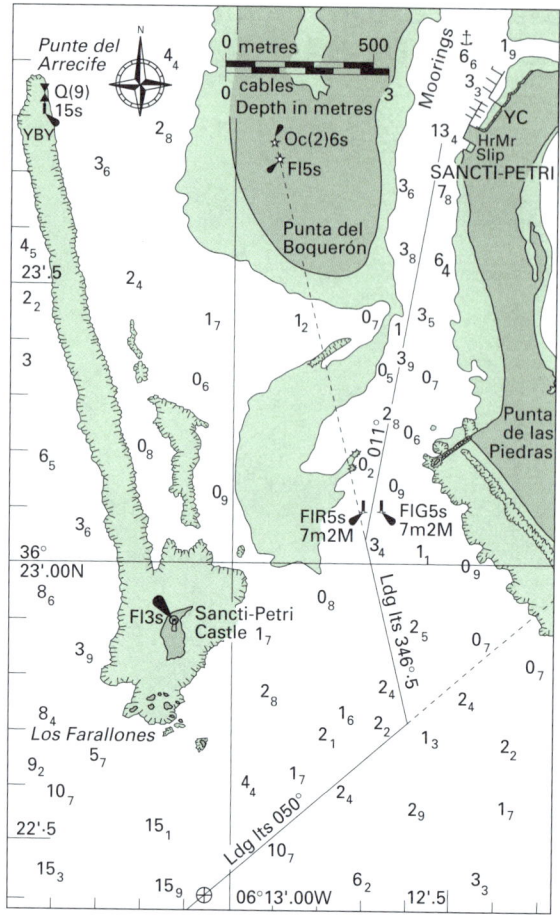

BARBATE IG 51

Cádiz 36°10'·89N 05°55'·50W

CHARTS
AC 142, 92; SC 4441, 444; SHOM 7042

TIDES
Standard Port LISBOA (←); ML 1·2; Zone –0100

Times				Height (metres)			
High Water		Low Water		MHWS	MHWN	MLWN	MLWS
0500	1000	0500	1100	3·8	3·0	1·4	0·5
1700	2200	1700	2300				
Differences CABO TRAFALGAR							
–0003	–0003	+0026	+0026	–1·4	–1·1	–0·5	–0·1
RIO BARBATE							
+0016	+0016	+0045	+0045	–1·9	–1·5	–0·4	+0·1
PUNTA CAMARINAL (36°05'N 05°48'W)							
–0007	–0007	+0013	+0013	–1·7	–1·4	–0·6	–0·2

SHELTER
Good. FVs berth in the large outer basin, yachts in the 3 inner basins (3m) to the W. Barbate (the "de Franco" is no longer used) is the most E'ly (outside the Med) of the new marinas run by Junta de Andalucía.

NAVIGATION
WPT 36°10'·50N 05°55'·20W, 117·5°/297·5° from/to SW mole hd, 2ca. Shoals (6·6m) extend SE along the coast. Tunny nets are a particular hazard, especially at night. A net is reported to extend 2M S from close off the bkwtr, April-Oct. Other nets may be laid W of Cabo Plata and NW of Tarifa.

LIGHTS AND MARKS
Barbate lt ho, Fl (2) WR 7s, W tr + R bands, is on edge of town 4ca N of hbr ent. Ldg lts, both Q, as chartlet, lead 297·5° between two pairs of lateral lt buoys: first pair, Q (2) R or G 5s; second pair, Q (3) R or G 7s. A 3rd pair, Q (4) R or G 12s, is close NE of marina ent.

RADIO TELEPHONE
Marina VHF Ch 09. Barbate is at W end of Gibraltar Strait VTS; monitor *Tarifa Traffic* Ch 10 16. See also IG 52.

TELEPHONE (Dial code 956)
Hr Mr, ⌗ and Met: via marina.

FACILITIES
Marina (421 AB, inc ⓥ), 1553ptas, ☎ 431907, ⚓ 431918, AC, P, FW, Slip, BH (32 ton), ME, El, Sh, Bar, R, Ice, ▫;
Club Náutico (portakabin). **Town:** V, R, Bar, Ⓑ, ✉; Bus to Cádiz (61km); ✈ Jerez, Gibraltar or Malaga.

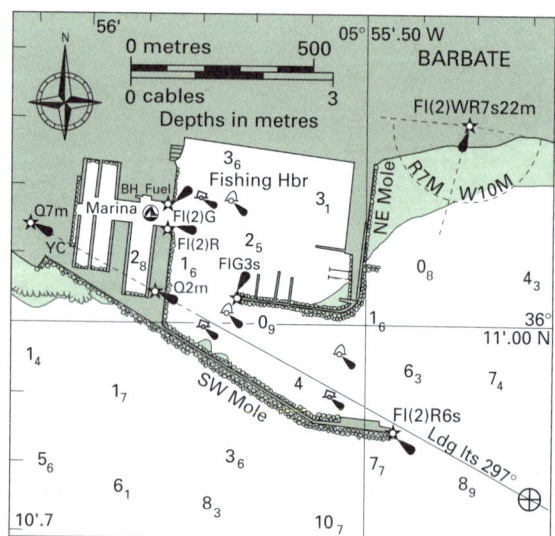

TIME ZONE –0100
(Gibraltar Standard Time)
Subtract 1 hour for UT
For Daylight Summer Time add
ONE hour in non-shaded areas

GIBRALTAR
LAT 36°08′N   LONG 5°21′W

TIMES AND HEIGHTS OF HIGH AND LOW WATERS

YEAR **1998**

Southwest Spain

	JANUARY				FEBRUARY				MARCH				APRIL		
	Time m		Time m		Time m		Time m		Time m		Time m		Time m		Time m
1 TH	0443 1.0 / 1018 0.1 / 1707 0.9 / 2244 0.0	**16** F	0526 0.8 / 1058 0.1 / 1742 0.8 / 2313 0.1	**1** SU	0603 1.0 / 1144 0.0 / 1830 0.9	**16** M	0603 0.8 / 1142 0.1 / 1823 0.7 / 2350 0.1	**1** SU	0501 1.0 / 1043 −0.1 / 1728 1.0 / 2302 −0.1	**16** M	0501 0.8 / 1042 0.1 / 1720 0.8 / 2251 0.0	**1** W	0615 0.9 / 1153 0.0 / 1848 0.9	**16** TH	0538 0.8 / 1118 0.1 / 1807 0.8 / 2327 0.2
2 F	0528 1.0 / 1104 0.1 / 1753 0.9 / 2330 0.1	**17** SA	0602 0.8 / 1135 0.1 / 1819 0.8 / 2348 0.1	**2** M	0005 0.0 / 0651 0.9 / 1235 0.1 / 1921 0.8	**17** TU	0639 0.7 / 1217 0.1 / 1901 0.7	**2** M	0546 1.0 / 1127 0.0 / 1815 0.9 / 2345 0.0	**17** TU	0531 0.8 / 1112 0.1 / 1754 0.8 / 2319 0.1	**2** TH	0009 0.1 / 0705 0.8 / 1242 0.1 / 1942 0.8	**17** F	0617 0.7 / 1155 0.2 / 1850 0.7
3 SA	0615 0.9 / 1155 0.1 / 1842 0.9	**18** SU	0639 0.7 / 1215 0.1 / 1858 0.7	**3** TU	0057 0.1 / 0745 0.8 / 1334 0.1 / 2019 0.8	**18** W	0024 0.1 / 0720 0.7 / 1259 0.2 / 1946 0.6	**3** TU	0634 0.9 / 1214 0.0 / 1905 0.8	**18** W	0605 0.8 / 1144 0.1 / 1831 0.7 / 2351 0.1	**3** F	0102 0.2 / 0803 0.7 / 1344 0.2 / 2044 0.7	**18** SA	0007 0.2 / 0703 0.7 / 1244 0.2 / 1942 0.7
4 SU	0021 0.1 / 0707 0.9 / 1253 0.2 / 1936 0.8	**19** M	0025 0.2 / 0719 0.7 / 1300 0.2 / 1940 0.7	**4** W	0159 0.2 / 0848 0.8 / 1450 0.2 / 2128 0.7	**19** TH	0106 0.2 / 0811 0.6 / 1356 0.2 / 2042 0.6	**4** W	0032 0.1 / 0725 0.8 / 1307 0.1 / 2000 0.8	**19** TH	0644 0.7 / 1220 0.2 / 1914 0.7	**4** SA	0214 0.3 / 0912 0.7 / 1523 0.3 / 2200 0.7	**19** SU	0101 0.3 / 0803 0.7 / 1359 0.3 / 2046 0.7
5 M	0121 0.2 / 0805 0.8 / 1402 0.2 / 2039 0.8	**20** TU	0109 0.2 / 0807 0.7 / 1354 0.2 / 2030 0.6	**5** F	0324 0.2 / 1001 0.7 / 1627 0.2 / 2250 0.7	**20** SA	0209 0.2 / 0907 0.6 / 1528 0.2 / 2153 0.6	**5** TH	0128 0.2 / 0825 0.8 / 1416 0.2 / 2105 0.7	**20** F	0029 0.2 / 0731 0.7 / 1310 0.2 / 2007 0.6	**5** SU	0405 0.7 / 1035 0.7 / 1705 0.7 / 2323 0.7	**20** M	0229 0.3 / 0921 0.6 / 1545 0.2 / 2204 0.7
6 TU	0236 0.2 / 0913 0.8 / 1523 0.2 / 2155 0.7	**21** W	0207 0.2 / 0904 0.6 / 1506 0.2 / 2132 0.6	**6** F	0457 0.2 / 1119 0.7 / 1749 0.2	**21** SA	0358 0.3 / 1034 0.6 / 1701 0.2 / 2314 0.6	**6** F	0247 0.2 / 0937 0.7 / 1603 0.2 / 2227 0.6	**21** SA	0124 0.2 / 0834 0.6 / 1433 0.2 / 2115 0.6	**6** M	0531 0.3 / 1154 0.7 / 1805 0.7	**21** TU	0416 0.3 / 1047 0.7 / 1701 0.2 / 2320 0.7
7 W	0401 0.2 / 1027 0.8 / 1645 0.2 / 2313 0.7	**22** TH	0333 0.3 / 1009 0.6 / 1627 0.2 / 2244 0.6	**7** SA	0010 0.7 / 0605 0.2 / 1228 0.7 / 1847 0.1	**22** SU	0523 0.2 / 1147 0.7 / 1803 0.1	**7** SA	0437 0.2 / 1100 0.7 / 1737 0.2 / 2353 0.7	**22** SU	0303 0.3 / 0955 0.6 / 1625 0.2 / 2238 0.6	**7** TU	0030 0.7 / 0623 0.2 / 1253 0.7 / 1847 0.2	**22** W	0528 0.2 / 1159 0.7 / 1757 0.1
8 TH	0515 0.2 / 1136 0.8 / 1754 0.1	**23** F	0455 0.3 / 1116 0.7 / 1733 0.2 / 2353 0.6	**8** SU	0114 0.7 / 0658 0.1 / 1325 0.8 / 1932 0.1	**23** M	0024 0.7 / 0618 0.1 / 1249 0.7 / 1851 0.1	**8** SU	0555 0.2 / 1217 0.7 / 1836 0.2	**23** M	0452 0.2 / 1118 0.6 / 1735 0.2 / 2354 0.7	**8** W	0120 0.8 / 0702 0.1 / 1337 0.8 / 1923 0.1	**23** TH	0023 0.8 / 0622 0.1 / 1256 0.8 / 1845 0.1
9 F	0022 0.8 / 0614 0.1 / 1238 0.9 / 1849 0.1	**24** SA	0553 0.2 / 1217 0.7 / 1824 0.1	**9** M	0204 0.8 / 0740 0.1 / 1413 0.8 / 2010 0.0	**24** TU	0119 0.7 / 0705 0.1 / 1340 0.8 / 1934 0.0	**9** M	0100 0.7 / 0646 0.2 / 1316 0.7 / 1918 0.1	**24** TU	0556 0.2 / 1226 0.7 / 1827 0.1	**9** TH	0159 0.8 / 0738 0.1 / 1414 0.8 / 1956 0.1	**24** F	0116 0.9 / 0710 0.1 / 1345 0.9 / 1930 0.0
10 SA	0120 0.8 / 0704 0.1 / 1331 0.9 / 1936 0.0	**25** SU	0051 0.7 / 0639 0.1 / 1310 0.8 / 1909 0.1	**10** TU	0245 0.8 / 0819 0.0 / 1454 0.9 / 2044 0.0	**25** W	0206 0.8 / 0749 0.0 / 1428 0.9 / 2016 −0.1	**10** TU	0149 0.8 / 0726 0.1 / 1400 0.8 / 1953 0.0	**25** W	0053 0.8 / 0646 0.1 / 1320 0.8 / 1912 0.0	**10** F	0232 0.8 / 0812 0.1 / 1447 0.8 / 2027 0.1	**25** SA	0203 1.0 / 0756 0.0 / 1432 0.9 / 2014 0.0
11 SU	0209 0.8 / 0748 0.1 / 1419 0.9 / 2017 0.0	**26** M	0140 0.8 / 0721 0.1 / 1358 0.8 / 1950 0.0	**11** W	0323 0.8 / 0854 0.0 / 1532 0.8 / 2117 0.0	**26** TH	0250 0.9 / 0832 0.0 / 1513 0.9 / 2058 −0.1	**11** W	0228 0.8 / 0802 0.0 / 1438 0.8 / 2025 0.0	**26** TH	0143 0.9 / 0731 0.0 / 1408 0.9 / 1955 0.0	**11** SA	0303 0.8 / 0844 0.1 / 1518 0.9 / 2058 0.1	**26** SU	0249 1.0 / 0840 −0.1 / 1518 1.0 / 2056 0.0
12 M	0253 0.8 / 0828 0.0 / 1503 0.9 / 2054 0.0	**27** TU	0224 0.9 / 0802 0.0 / 1443 0.9 / 2031 0.0	**12** TH	0357 0.8 / 0929 0.0 / 1608 0.8 / 2148 0.0	**27** F	0334 1.0 / 0915 −0.1 / 1558 1.0 / 2139 −0.1	**12** TH	0302 0.8 / 0836 0.0 / 1513 0.8 / 2055 0.0	**27** F	0229 0.9 / 0816 −0.1 / 1454 0.9 / 2037 −0.1	**12** SU	0332 0.9 / 0916 0.1 / 1549 0.9 / 2127 0.1	**27** M	0334 1.0 / 0924 −0.1 / 1605 1.0 / 2139 0.0
13 TU	0334 0.9 / 0906 0.0 / 1546 0.9 / 2130 0.0	**28** W	0307 0.9 / 0844 0.0 / 1528 0.9 / 2112 −0.1	**13** F	0429 0.8 / 1003 0.0 / 1642 0.8 / 2219 0.0	**28** SA	0417 1.0 / 0959 −0.1 / 1642 1.0 / 2220 −0.1	**13** F	0333 0.9 / 0909 0.0 / 1545 0.9 / 2125 0.0	**28** SA	0313 1.0 / 0859 −0.1 / 1539 1.0 / 2119 −0.1	**13** M	0401 0.9 / 0947 0.0 / 1621 0.9 / 2155 0.1	**28** TU	0420 1.0 / 1006 −0.1 / 1652 1.0 / 2221 0.0
14 W	0413 0.9 / 0944 0.0 / 1626 0.9 / 2205 0.0	**29** TH	0350 1.0 / 0927 0.0 / 1612 0.9 / 2153 −0.1	**14** SA	0500 0.8 / 1036 0.0 / 1716 0.8 / 2249 0.0			**14** SA	0403 0.8 / 0941 0.0 / 1617 0.8 / 2154 0.0	**29** SU	0357 1.0 / 0942 −0.1 / 1624 1.0 / 2200 −0.1	**14** TU	0432 0.8 / 1016 0.0 / 1654 0.8 / 2224 0.1	**29** W	0507 1.0 / 1049 0.0 / 1740 0.9 / 2304 0.1
15 TH	0450 0.8 / 1021 0.0 / 1705 0.8 / 2239 0.0	**30** F	0433 1.0 / 1011 0.0 / 1657 1.0 / 2236 −0.1	**15** SU	0531 0.8 / 1109 0.0 / 1749 0.8 / 2319 0.0			**15** SU	0431 0.8 / 1012 0.0 / 1648 0.8 / 2223 0.0	**30** M	0442 1.0 / 1025 −0.1 / 1711 1.0 / 2242 0.0	**15** W	0504 0.8 / 1046 0.1 / 1729 0.8 / 2253 0.1	**30** TH	0556 0.9 / 1132 0.1 / 1830 0.9 / 2349 0.1
		31 SA	0517 1.0 / 1056 0.0 / 1743 0.9 / 2319 0.0							**31** TU	0527 1.0 / 1108 0.0 / 1758 0.9 / 2324 0.0				

Chart Datum: 0·25 metres below Alicante Datum (Mean Sea Level, Alicante)

Harbour, Coastal and Tidal Information

For GMT: −1
For Spain: +1   (UT+2)

TIME ZONE −0100
(Gibraltar Standard Time)
Subtract 1 hour for UT
For Daylight Summer Time add
ONE hour in non-shaded areas

GIBRALTAR
LAT 36°08′N   LONG 5°21′W

TIMES AND HEIGHTS OF HIGH AND LOW WATERS

YEAR 1998

MAY

Day	Time	m	Day	Time	m
1 F	0647 / 1219 / 1923	0.8 / 0.1 / 0.8	**16** SA	0600 / 1139 / 1831 / 2354	0.8 / 0.2 / 0.8 / 0.2
2 SA	0040 / 0741 / 1315 / 2021	0.2 / 0.8 / 0.2 / 0.8	**17** SU	0646 / 1228 / 1922	0.8 / 0.2 / 0.8
3 SU	0144 / 0845 / 1431 / 2127	0.3 / 0.7 / 0.2 / 0.7	**18** M	0049 / 0742 / 1335 / 2021	0.3 / 0.7 / 0.2 / 0.8
4 M	0313 / 0958 / 1605 / 2239	0.3 / 0.7 / 0.3 / 0.7	**19** TU	0205 / 0853 / 1503 / 2131	0.3 / 0.7 / 0.2 / 0.8
5 TU	0444 / 1113 / 1714 / 2345	0.3 / 0.7 / 0.3 / 0.7	**20** W	0337 / 1014 / 1623 / 2245	0.3 / 0.7 / 0.2 / 0.8
6 W	0544 / 1214 / 1804	0.2 / 0.7 / 0.2	**21** TH	0455 / 1129 / 1725 / 2350	0.2 / 0.8 / 0.1 / 0.8
7 TH	0037 / 0628 / 1302 / 1844	0.8 / 0.2 / 0.7 / 0.2	**22** F	0556 / 1229 / 1818	0.1 / 0.8 / 0.1
8 F	0119 / 0707 / 1341 / 1921	0.8 / 0.2 / 0.8 / 0.2	**23** SA	0046 / 0649 / 1322 / 1906	0.9 / 0.1 / 0.9 / 0.1
9 SA	0155 / 0743 / 1416 / 1955	0.8 / 0.1 / 0.8 / 0.1	**24** SU	0137 / 0737 / 1411 / 1952	0.9 / 0.1 / 0.9 / 0.0
10 SU ○	0228 / 0817 / 1449 / 2028	0.8 / 0.1 / 0.8 / 0.1	**25** M ●	0226 / 0822 / 1459 / 2036	1.0 / 0.0 / 0.9 / 0.0
11 M ○	0300 / 0850 / 1522 / 2059	0.9 / 0.1 / 0.9 / 0.1	**26** TU	0313 / 0906 / 1546 / 2120	1.0 / 0.0 / 1.0 / 0.0
12 TU	0333 / 0922 / 1556 / 2129	0.9 / 0.1 / 0.9 / 0.1	**27** W	0400 / 0949 / 1634 / 2203	1.0 / 0.0 / 0.9 / 0.0
13 W	0407 / 0953 / 1631 / 2200	0.9 / 0.1 / 0.9 / 0.1	**28** TH	0448 / 1031 / 1722 / 2246	0.9 / 0.0 / 0.9 / 0.1
14 TH	0442 / 1025 / 1708 / 2233	0.9 / 0.1 / 0.9 / 0.2	**29** F	0537 / 1113 / 1811 / 2331	0.9 / 0.1 / 0.9 / 0.1
15 F	0519 / 1100 / 1747 / 2310	0.8 / 0.1 / 0.8 / 0.2	**30** SA	0626 / 1157 / 1900	0.8 / 0.1 / 0.8
			31 SU	0018 / 0717 / 1246 / 1951	0.2 / 0.8 / 0.2 / 0.8

JUNE

Day	Time	m	Day	Time	m
1 M	0113 / 0812 / 1343 / 2047	0.3 / 0.7 / 0.2 / 0.7	**16** TU	0037 / 0728 / 1313 / 1957	0.2 / 0.8 / 0.2 / 0.8
2 TU	0219 / 0913 / 1455 / 2147	0.3 / 0.7 / 0.2 / 0.7	**17** W	0143 / 0829 / 1424 / 2101	0.2 / 0.8 / 0.2 / 0.8
3 W	0337 / 1018 / 1610 / 2247	0.3 / 0.7 / 0.3 / 0.7	**18** TH	0301 / 0943 / 1543 / 2211	0.2 / 0.8 / 0.2 / 0.8
4 TH	0449 / 1122 / 1712 / 2343	0.3 / 0.7 / 0.3 / 0.7	**19** F	0422 / 1058 / 1653 / 2319	0.2 / 0.8 / 0.2 / 0.8
5 F	0545 / 1216 / 1802	0.3 / 0.7 / 0.2	**20** SA	0532 / 1204 / 1754	0.2 / 0.8 / 0.2
6 SA	0031 / 0631 / 1302 / 1844	0.8 / 0.2 / 0.8 / 0.2	**21** SU	0020 / 0630 / 1302 / 1846	0.9 / 0.1 / 0.8 / 0.1
7 SU	0113 / 0711 / 1342 / 1922	0.8 / 0.2 / 0.8 / 0.2	**22** M	0115 / 0722 / 1354 / 1935	0.9 / 0.1 / 0.9 / 0.1
8 M	0152 / 0748 / 1419 / 1958	0.8 / 0.1 / 0.8 / 0.2	**23** TU	0206 / 0809 / 1443 / 2021	0.9 / 0.0 / 0.9 / 0.0
9 TU	0229 / 0823 / 1456 / 2032	0.9 / 0.1 / 0.9 / 0.1	**24** W ●	0255 / 0852 / 1531 / 2104	0.9 / 0.0 / 0.9 / 0.1
10 W ○	0307 / 0857 / 1532 / 2106	0.9 / 0.1 / 0.9 / 0.2	**25** TH	0343 / 0933 / 1617 / 2147	0.9 / 0.0 / 0.9 / 0.1
11 TH	0345 / 0932 / 1610 / 2141	0.9 / 0.1 / 0.9 / 0.2	**26** F	0430 / 1013 / 1702 / 2228	0.9 / 0.1 / 0.9 / 0.1
12 F	0423 / 1007 / 1649 / 2218	0.9 / 0.1 / 0.9 / 0.2	**27** SA	0516 / 1052 / 1747 / 2310	0.9 / 0.1 / 0.9 / 0.1
13 SA	0504 / 1045 / 1730 / 2259	0.9 / 0.1 / 0.9 / 0.2	**28** SU	0601 / 1132 / 1831 / 2353	0.9 / 0.1 / 0.9 / 0.2
14 SU	0546 / 1127 / 1813 / 2344	0.9 / 0.1 / 0.9 / 0.2	**29** M	0647 / 1213 / 1915	0.8 / 0.1 / 0.8
15 M	0633 / 1214 / 1902	0.8 / 0.1 / 0.9	**30** TU	0039 / 0733 / 1259 / 2000	0.2 / 0.8 / 0.2 / 0.8

JULY

Day	Time	m	Day	Time	m
1 W	0130 / 0823 / 1352 / 2050	0.3 / 0.7 / 0.3 / 0.7	**16** TH	0119 / 0806 / 1350 / 2032	0.2 / 0.8 / 0.2 / 0.9
2 TH	0230 / 0919 / 1458 / 2145	0.3 / 0.7 / 0.3 / 0.7	**17** F	0228 / 0915 / 1505 / 2140	0.2 / 0.8 / 0.2 / 0.9
3 F	0342 / 1021 / 1611 / 2243	0.3 / 0.7 / 0.3 / 0.7	**18** SA	0351 / 1031 / 1625 / 2252	0.2 / 0.8 / 0.3 / 0.8
4 SA	0453 / 1123 / 1715 / 2339	0.3 / 0.7 / 0.3 / 0.7	**19** SU	0514 / 1144 / 1736	0.2 / 0.8 / 0.2
5 SU	0551 / 1219 / 1807	0.2 / 0.7 / 0.3	**20** M	0000 / 0619 / 1248 / 1833	0.9 / 0.2 / 0.9 / 0.2
6 M	0031 / 0638 / 1308 / 1850	0.8 / 0.2 / 0.8 / 0.2	**21** TU	0100 / 0712 / 1343 / 1923	0.9 / 0.1 / 0.9 / 0.1
7 TU	0118 / 0719 / 1350 / 1929	0.8 / 0.2 / 0.8 / 0.2	**22** W	0153 / 0757 / 1432 / 2008	0.9 / 0.1 / 0.9 / 0.1
8 W	0201 / 0757 / 1430 / 2007	0.9 / 0.1 / 0.9 / 0.2	**23** TH ●	0242 / 0838 / 1516 / 2050	0.9 / 0.1 / 0.9 / 0.1
9 TH	0243 / 0834 / 1510 / 2045	0.9 / 0.1 / 0.9 / 0.1	**24** F	0327 / 0916 / 1558 / 2129	0.9 / 0.0 / 0.9 / 0.1
10 F ○	0324 / 0911 / 1549 / 2123	0.9 / 0.1 / 1.0 / 0.1	**25** SA	0410 / 0952 / 1639 / 2208	0.9 / 0.1 / 0.9 / 0.1
11 SA	0406 / 0949 / 1630 / 2204	0.9 / 0.1 / 0.9 / 0.1	**26** SU	0451 / 1027 / 1717 / 2245	0.9 / 0.1 / 0.9 / 0.2
12 SU	0448 / 1029 / 1711 / 2246	1.0 / 0.1 / 1.0 / 0.1	**27** M	0530 / 1102 / 1754 / 2323	1.0 / 0.1 / 0.9 / 0.2
13 M	0531 / 1111 / 1755 / 2331	1.0 / 0.1 / 1.0 / 0.1	**28** TU	0610 / 1137 / 1831	0.8 / 0.2 / 0.9
14 TU	0618 / 1156 / 1841	0.9 / 0.1 / 1.0	**29** W	0002 / 0649 / 1215 / 1909	0.2 / 0.8 / 0.2 / 0.8
15 W	0021 / 0708 / 1248 / 1933	0.2 / 0.9 / 0.1 / 0.9	**30** TH	0043 / 0732 / 1256 / 1951	0.2 / 0.8 / 0.2 / 0.8
			31 F	0130 / 0821 / 1347 / 2041	0.3 / 0.7 / 0.3 / 0.7

AUGUST

Day	Time	m	Day	Time	m
1 SA	0232 / 0919 / 1459 / 2140	0.3 / 0.7 / 0.3 / 0.7	**16** SU	0325 / 1008 / 1602 / 2229	0.3 / 0.8 / 0.3 / 0.8
2 SU	0353 / 1026 / 1623 / 2246	0.3 / 0.7 / 0.3 / 0.7	**17** M	0502 / 1129 / 1723 / 2346	0.3 / 0.8 / 0.3 / 0.8
3 M	0510 / 1134 / 1730 / 2350	0.3 / 0.7 / 0.3 / 0.8	**18** TU	0612 / 1238 / 1824	0.2 / 0.8 / 0.2
4 TU	0606 / 1233 / 1820	0.3 / 0.8 / 0.3	**19** W	0051 / 0702 / 1333 / 1911	0.9 / 0.2 / 0.9 / 0.2
5 W	0045 / 0651 / 1322 / 1903	0.8 / 0.2 / 0.8 / 0.2	**20** TH	0144 / 0743 / 1419 / 1953	0.9 / 0.1 / 0.9 / 0.1
6 TH	0135 / 0731 / 1405 / 1944	0.9 / 0.1 / 0.9 / 0.2	**21** F	0228 / 0819 / 1459 / 2031	0.9 / 0.1 / 1.0 / 0.1
7 F	0220 / 0810 / 1447 / 2024	0.9 / 0.1 / 1.0 / 0.1	**22** SA ●	0308 / 0853 / 1535 / 2107	0.9 / 0.1 / 1.0 / 0.1
8 SA	0303 / 0848 / 1528 / 2105	1.0 / 0.1 / 1.0 / 0.1	**23** SU	0345 / 0925 / 1610 / 2142	1.0 / 0.1 / 1.0 / 0.1
9 SU ○	0346 / 0928 / 1609 / 2146	1.0 / 0.0 / 1.1 / 0.1	**24** M	0420 / 0957 / 1642 / 2217	0.9 / 0.1 / 1.0 / 0.1
10 M	0429 / 1008 / 1651 / 2229	1.0 / 0.0 / 1.1 / 0.1	**25** TU	0454 / 1028 / 1714 / 2250	0.9 / 0.1 / 0.9 / 0.1
11 TU	0513 / 1050 / 1734 / 2314	1.0 / 0.1 / 1.1 / 0.1	**26** W	0529 / 1100 / 1745 / 2323	0.9 / 0.2 / 0.9 / 0.2
12 W	0559 / 1134 / 1820	0.9 / 0.1 / 1.0	**27** TH	0604 / 1132 / 1818 / 2358	0.9 / 0.2 / 0.9 / 0.2
13 TH	0001 / 0649 / 1222 / 1909	0.1 / 1.0 / 0.2 / 1.0	**28** F	0642 / 1206 / 1856	0.8 / 0.3 / 0.8
14 F	0054 / 0744 / 1318 / 2005	0.2 / 0.9 / 0.2 / 0.9	**29** SA	0037 / 0727 / 1248 / 1941	0.3 / 0.8 / 0.3 / 0.8
15 SA	0157 / 0850 / 1430 / 2112	0.3 / 0.8 / 0.3 / 0.9	**30** SU	0127 / 0823 / 1346 / 2040	0.3 / 0.7 / 0.4 / 0.7
			31 M	0248 / 0932 / 1524 / 2153	0.4 / 0.7 / 0.4 / 0.7

Chart Datum: 0·25 metres below Alicante Datum (Mean Sea Level, Alicante)

TIME ZONE –0100
(Gibraltar Standard Time)
Subtract 1 hour for UT
For Gibraltar Summer Time add ONE hour in non-shaded areas

GIBRALTAR

LAT 36°08′N   LONG 5°21′W

TIMES AND HEIGHTS OF HIGH AND LOW WATERS

YEAR **1998**

Chart Datum: 0·25 metres below Alicante Datum (Mean Sea Level, Alicante)

SEPTEMBER				OCTOBER				NOVEMBER				DECEMBER			
Time	m	Time	m	Time	m	Time	m	Time	m	Time	m	Time	m	Time	m
1 0430 1049 TU 1653 2311	0.3 0.7 0.4 0.8	**16** 0559 1225 W 1810	0.3 0.9 0.3	**1** 0505 1122 TH 1723 2348	0.3 0.8 0.3 0.8	**16** 0024 0617 F 1251 1830	0.8 0.3 0.9 0.3	**1** 0019 0612 SU 1242 1833	0.9 0.2 1.0 0.1	**16** 0119 0658 M 1332 1915	0.9 0.2 0.9 0.2	**1** 0048 0634 TU 1305 1900	0.9 0.1 1.0 0.0	**16** 0125 0707 W 1336 1927	0.8 0.2 0.8 0.1
2 0536 1158 W 1752	0.3 0.8 0.3	**17** 0042 0645 TH 1318 1854	0.9 0.3 0.9 0.2	**2** 0556 1221 F 1812	0.2 0.9 0.2	**17** 0111 0654 SA 1331 1907	0.9 0.3 1.0 0.2	**2** 0110 0655 M 1329 1917	1.0 0.1 1.1 0.1	**17** 0152 0731 TU 1404 1949	0.9 0.2 0.9 0.1	**2** 0138 0720 W 1354 1946	1.0 0.1 1.0 0.0	**17** 0201 0741 TH 1412 2003	0.8 0.2 0.9 0.1
3 0016 0623 TH 1253 1838	0.8 0.2 0.9 0.2	**18** 0131 0722 F 1359 1933	0.9 0.2 1.0 0.2	**3** 0045 0638 SA 1311 1856	0.9 0.2 1.0 0.1	**18** 0147 0727 SU 1405 1941	0.9 0.2 1.0 0.1	**3** 0157 0737 TU 1415 2001	1.0 0.1 1.1 0.0	**18** 0223 0802 W 1435 2022	0.9 0.2 0.9 0.1	**3** 0226 0804 TH 1442 2031	1.0 0.0 1.1 0.0	**18** 0235 0814 F 1448 ● 2037	0.9 0.2 0.9 0.1
4 0110 0705 F 1339 1920	0.9 0.2 1.0 0.2	**19** 0210 0755 SA 1435 2008	0.9 0.2 1.0 0.1	**4** 0133 0719 SU 1355 1939	1.0 0.1 1.1 0.1	**19** 0219 0758 M 1435 2015	0.9 0.2 1.0 0.1	**4** 0242 0819 W ○ 1500 2045	0.9 0.0 1.1 0.0	**19** 0253 0833 TH 1507 ● 2054	0.9 0.2 1.0 0.1	**4** 0312 0848 F 1529 2115	1.0 0.0 1.0 0.0	**19** 0309 0847 SA 1524 2110	0.9 0.2 0.9 0.1
5 0157 0744 SA 1422 2002	1.0 0.1 1.0 0.1	**20** 0245 0826 SU 1507 ● 2042	1.0 0.1 1.0 0.1	**5** 0218 0759 M 1439 ○ 2022	1.1 0.1 1.1 0.0	**20** 0248 0826 TU 1504 ● 2047	1.0 0.2 1.0 0.1	**5** 0327 0901 TH 1545 2128	1.1 0.0 1.1 0.0	**20** 0324 0903 F 1539 2125	1.0 0.2 0.9 0.1	**5** 0359 0932 SA 1617 2158	1.0 0.1 1.0 0.0	**20** 0343 0920 SU 1601 2144	0.9 0.1 0.9 0.1
6 0241 0824 SU 1504 ○ 2044	1.0 0.1 1.1 0.1	**21** 0317 0856 M 1537 2114	1.0 0.1 1.0 0.1	**6** 0302 0840 TU 1522 2104	1.1 0.0 1.2 0.0	**21** 0317 0857 W 1533 2117	1.0 0.2 1.0 0.1	**6** 0413 0944 F 1632 2211	1.1 0.1 1.1 0.1	**21** 0357 0933 SA 1613 2156	0.9 0.2 0.9 0.1	**6** 0447 1016 SU 1706 2241	1.0 0.1 1.0 0.1	**21** 0420 0954 M 1639 2219	0.9 0.2 0.9 0.1
7 0324 0904 M 1546 2126	1.1 0.0 1.1 0.0	**22** 0348 0926 TU 1606 2146	1.0 0.1 1.0 0.1	**7** 0346 0921 W 1606 2147	1.1 0.0 1.2 0.0	**22** 0347 0926 TH 1602 2147	1.0 0.2 1.0 0.1	**7** 0501 1028 SA 1719 2255	1.0 0.1 1.0 0.1	**22** 0432 1005 SU 1648 2229	0.9 0.2 0.9 0.2	**7** 0535 1103 M 1755 2326	0.9 0.1 0.9 0.1	**22** 0458 1032 TU 1718 2256	0.9 0.2 0.9 0.1
8 0408 0945 TU 1629 2209	1.1 0.0 1.2 0.0	**23** 0418 0955 W 1634 2217	1.0 0.1 1.0 0.1	**8** 0432 1003 TH 1651 2230	1.1 0.1 1.1 0.1	**23** 0418 0955 F 1632 2217	1.0 0.2 1.0 0.2	**8** 0550 1115 SU 1810 2341	0.9 0.2 1.0 0.2	**23** 0510 1040 M 1726 2305	0.9 0.2 0.9 0.2	**8** 0625 1152 TU 1846	0.9 0.2 0.8	**23** 0540 1114 W 1801 2339	0.9 0.2 0.8 0.2
9 0453 1026 W 1712 2252	1.1 0.0 1.1 0.1	**24** 0449 1024 TH 1704 2247	0.9 0.2 0.9 0.2	**9** 0519 1047 F 1737 2314	1.1 0.1 1.1 0.1	**24** 0451 1025 SA 1705 2248	0.9 0.2 0.9 0.2	**9** 0644 1207 M 1904	0.9 0.2 0.9	**24** 0552 1121 TU 1809 2347	0.9 0.2 0.8 0.3	**9** 0014 0718 W 1248 1939	0.2 0.8 0.3 0.8	**24** 0626 1203 TH 1848	0.9 0.2 0.8
10 0539 1110 TH 1758 2337	1.1 0.1 1.1 0.1	**25** 0522 1054 F 1735 2318	0.9 0.2 0.9 0.2	**10** 0608 1133 SA 1826	1.0 0.2 1.0	**25** 0528 1058 SU 1741 2322	0.9 0.3 0.9 0.3	**10** 0035 0743 TU 1312 2005	0.3 0.9 0.4 0.8	**25** 0641 1212 W 1900	0.8 0.3 0.8	**10** 0111 0816 TH 1354 2039	0.3 0.8 0.4 0.7	**25** 0029 0717 F 1303 1944	0.2 0.8 0.3 0.8
11 0628 1156 F 1847	1.0 0.2 1.0	**26** 0559 1126 SA 1811 2353	0.9 0.3 0.9 0.3	**11** 0002 0702 SU 1226 1921	0.2 0.9 0.3 0.9	**26** 0612 1136 M 1824	0.9 0.3 0.8	**11** 0148 0853 W 1438 2119	0.4 0.8 0.4 0.8	**26** 0043 0738 TH 1323 2003	0.3 0.8 0.3 0.8	**11** 0223 0920 F 1511 2147	0.3 0.8 0.3 0.7	**26** 0135 0819 SA 1418 2052	0.2 0.8 0.2 0.7
12 0027 0723 SA 1250 1941	0.2 0.9 0.3 0.9	**27** 0643 1204 SU 1855	0.8 0.3 0.8	**12** 0100 0806 M 1335 2026	0.3 0.9 0.4 0.8	**27** 0004 0703 TU 1227 1918	0.3 0.8 0.4 0.8	**12** 0331 1010 TH 1609 2242	0.4 0.8 0.4 0.8	**27** 0211 0847 F 1456 2121	0.3 0.8 0.3 0.7	**12** 0347 1026 SA 1624 2258	0.3 0.8 0.3 0.7	**27** 0258 0929 SU 1541 2210	0.2 0.8 0.2 0.7
13 0128 0827 SU 1400 2048	0.3 0.9 0.4 0.9	**28** 0037 0737 M 1256 1951	0.3 0.8 0.4 0.8	**13** 0227 0923 TU 1516 2150	0.4 0.8 0.4 0.8	**28** 0106 0807 W 1349 2028	0.4 0.8 0.4 0.8	**13** 0450 1120 F 1712 2351	0.4 0.8 0.3 0.8	**28** 0346 1004 SA 1617 2244	0.3 0.8 0.3 0.8	**13** 0455 1125 SU 1722 2358	0.3 0.8 0.3 0.7	**28** 0420 1043 M 1655 2327	0.2 0.8 0.2 0.8
14 0259 0947 M 1541 2211	0.4 0.8 0.4 0.8	**29** 0148 0845 TU 1427 2105	0.4 0.8 0.4 0.7	**14** 0424 1049 W 1649 2319	0.4 0.8 0.4 0.8	**29** 0300 0930 TH 1538 2156	0.4 0.9 0.4 0.8	**14** 0542 1213 SA 1759	0.3 0.9 0.3	**29** 0454 1114 SU 1719 2353	0.2 0.9 0.2 0.8	**14** 0547 1215 M 1809	0.3 0.8 0.3	**29** 0525 1149 TU 1758	0.2 0.9 0.1
15 0451 1114 TU 1711 2336	0.4 0.8 0.4 0.8	**30** 0349 1005 W 1617 2233	0.4 0.8 0.4 0.8	**15** 0533 1200 TH 1747	0.4 0.9 0.3	**30** 0430 1043 F 1652 2318	0.3 0.8 0.3 0.8	**15** 0040 0622 SU 1256 1838	0.8 0.3 0.9 0.2	**30** 0547 1213 M 1812	0.2 0.9 0.1	**15** 0046 0629 TU 1257 1850	0.8 0.2 0.8 0.2	**30** 0031 0620 W 1247 1852	0.8 0.1 0.9 0.0
						31 0526 1148 SA 1746	0.3 0.9 0.2							**31** 0126 0710 TH 1340 1941	0.9 0.1 0.9 0.0

STRAIT OF GIBRALTAR

IG 52

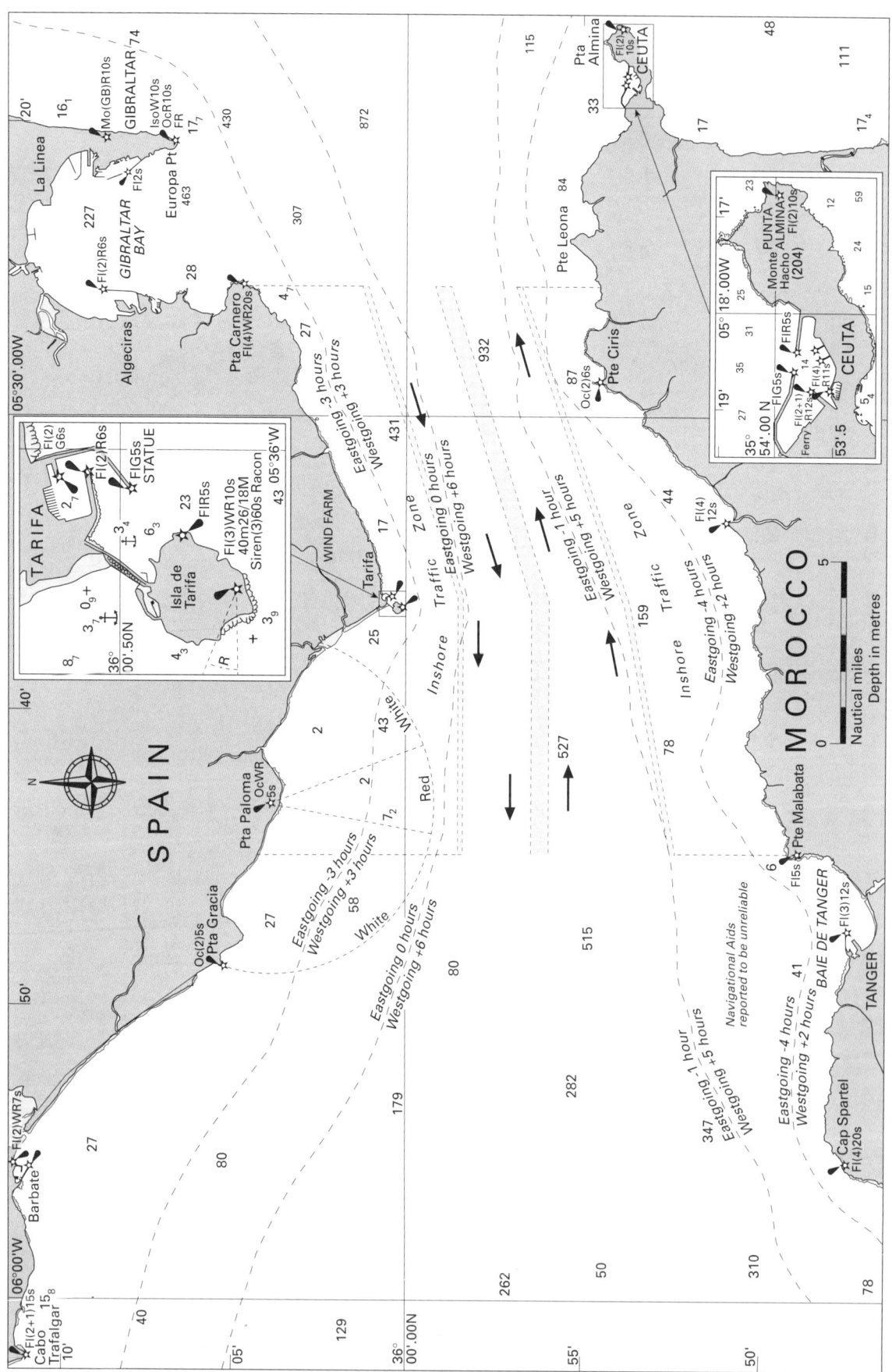

Southwest Spain

MINOR HARBOURS ON THE GIBRALTAR STRAIT

TARIFA 36°00'·40N 05°36'·20W. AC 142; SC 4450, 445B; SHOM 1619, 7042. Inset chartlet is opposite. Tides, see IG 53. A FV, naval and ferry hbr with no yacht berths. Ferries berth on the outer end of the outer mole and on S side of the hbr; FVs fill the S and W sides of the hbr; naval pens and *Guardia Civil* craft use the N side. Best options for yachts are the N end of outer mole or, by agreement, the naval pens. ⚓ off hbr ent in 5m; if a *levanter* is blowing ⚓ NW of Isla de Tarifa (a prohib military area joined by causeway to mainland). Lts: Lt ho, Fl (3) WR 10s 40m 26/18M, W113-089°, R089°-113°; Siren (3) 60s; Racon, RDF bn (see IG 27). E side of Isla de Tarifa, Fl R 5s 12m 3M. Outer mole hd, Fl G 5s 10m 5M. Inner mole hd, Fl (2) R 6s 6m 1M. Inner spur Fl (2) G 6s 5m 2M.

Tarifa VTS, c/s *Tarifa Trafico* VHF Ch **10** 74 16, provides a reporting service (voluntary for yachts). Its area covers the Strait and TSS/ITZ between 05°58'W and 05°15'W. The service provides: listening watch for distress H24; broadcasts in Spanish and English at every even H+15 of traffic, nav and weather info in the area; and radar info on request. Tarifa VTS is part of Tarifa MRCC which also broadcasts Navtex info (ident G).

CEUTA 35°53'·82N 05°18'·50W. AC 2742, 142, 3578; SC 4511; SHOM 7503. Inset chartlet, see opposite. Tides, see IG 54. Ceuta, a Spanish enclave in Morocco, is only 13M S of Europa Pt. It lies on the isthmus between 848m high mountains to the W and the Peninsula de Almina to the E, on which Monte Hacho (Pillars of Hercules) fortress is prominent. Pta Almina lt ho at the E tip is conspic, Fl (2) 10s 148m 22M, siren (2) 45s. The hbr is exposed to ENE winds (*levanter*), but the new marina is well sheltered. Fuel is available at the marina.

SURFACE FLOW IN THE GIBRALTAR STRAIT

Surface flow is the net effect of the prevailing E-going current from the Atlantic into the Med (further influenced by recent E or W winds) and tidal streams. Max surface flows are about 2kn W-going and 5kn E-going. Max surface current is about 2kn W-going and 5kn E-going. Max surface current is about 2kn E-going in the middle of the Strait. Max tidal stream rates are about 3kn in either direction, usually to be found close inshore around headlands. Thus as surface current decreases away from mid-Strait, tidal streams increase closer inshore.

Tidal streams are insufficiently observed and are therefore approximate, although generally accurate. Near the middle of the Strait the E-going stream starts at HW Gib and the W-going at HWG+6, as depicted on the chartlet opposite by a dashed line nearly coincident with the N side of the TSS. Closer inshore, streams start to make progressively earlier, the times shown by extra dashed lines.

Tidal races on the Spanish side form: off Cape Trafalgar, usually up to 2M SW, but 5-12M in heavy weather; near Bajo de Los Cabezos (2-4M S of Pta Paloma); and off Tarifa. As ever the position and violence of these races depends on wind, especially over tide, springs/neaps and the state of the tide.

****

Nobly, nobly Cape Saint Vincent to the North-west died away;
Sunset ran, one glorious blood-red, reeking into Cádiz Bay;
Bluish 'mid the burning water, full in face Trafalgar lay;
In the dimmest North-east distance dawned Gibraltar grand and gray;
'Here and here did England help me: how can I help England?' – say,
Whoso turns as I, this evening, turn to God to praise and pray,
While Jove's planet rises yonder, silent over Africa.

Home Thoughts from the Sea: Robert Browning

ALGECIRAS
Cádiz 36°07'·20N 05°26'·15W

IG 53

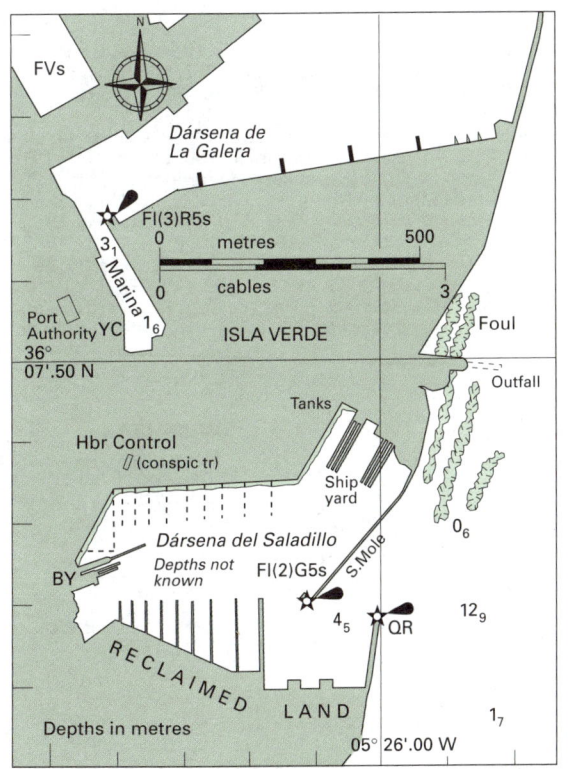

CHARTS
AC 1448, 142, 3578; SC 4451, 445A, 445; SHOM 7026, 7042
TIDES
Standard Port GIBRALTAR (→); ML 0·66; Zone –0100

Times				Height (metres)			
High Water		Low Water		MHWS	MHWN	MLWN	MLWS
0000	0700	0100	0600	1·0	0·7	0·3	0·1
1200	1900	1300	1800				
Differences TARIFA							
–0038	–0038	–0042	–0042	+0·4	+0·3	+0·3	+0·2
PUNTA CARNERO							
–0010	–0010	0000	0000	0·0	+0·1	+0·1	+0·1
ALGECIRAS							
–0010	–0010	–0010	–0010	+0·1	+0·2	+0·1	+0·1

SHELTER
Good, except perhaps in strong SE'lies. A new marina at Darsena del Saladillo, S of the docks, should partly open in 1998; check with Port Authority before visiting. It may be possible to ⚓ in NE corner, clear of WIP. Existing small marina/YC will shift to new marina in 1998.
NAVIGATION
WPT 36°06'·80N 05°24'·68W, ECM buoy, Q (3) 10s, (off chartlet) 109°/289° from/to new marina ent, 1·25M. From S, keep 5ca off Pta de San García. From N/E, beware drying reefs close N of ent; also big ships at ⚓ in the bay.
LIGHTS AND MARKS
Conspic Hbr Control twr leads 289° by day from WPT to new marina.
RADIO TELEPHONE
Marina VHF Ch 09 16. Port Ch 09, 12, 13, 16.
TELEPHONE (Dial code 956)
Hr Mr 572620; Port Authority 585400, ☎ 585445; ⌗ and Met via marina; Brit Vice-Consul 661600/04.
FACILITIES
Marina (existing) (little or no space for Ⓥ), ☎ 572503, AC, D, FW, ME, Slip; **Real Club Náutico** ☎ 572503, R, Bar;
New marina: no details available; contact Port Authority.
Town: V, R, Bar, Ⓑ, ✉, ⇌; ✈ Jerez, Gibraltar, Malaga.

GIBRALTAR   IG 54

36°08'·50N 05°22'·00W

CHARTS
AC 45, 144, 1448, 142; SC 4451/2, 445A; SHOM 7026, 7042

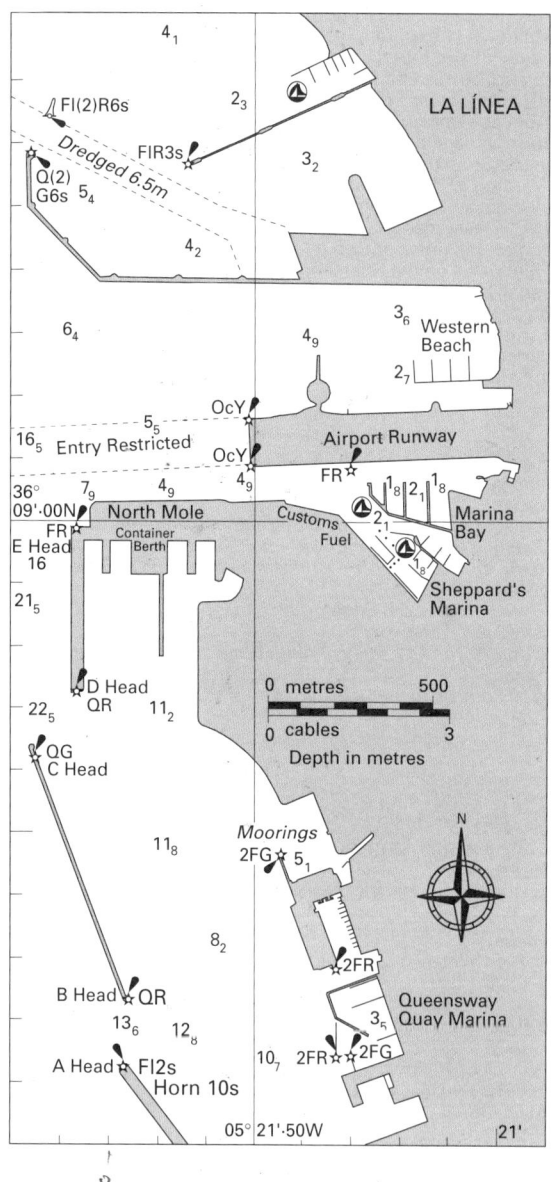

TIDES
ML 0·48; Zone –0100. Gibraltar is a Standard Port. Daily predictions are given below.

Standard Port GIBRALTAR (→)

Times				Height (metres)			
High Water		Low Water		MHWS	MHWN	MLWN	MLWS
0000	0700	0100	0600	1·0	0·7	0·3	0·1
1200	1900	1300	1800				
Differences SANDY BAY (E side of the Rock)							
–0011	–0011	–0016	–0016	–0·1	–0·1	0·0	0·0
TANGER (Morocco)							
–0010	–0010	–0050	+0010	+1·4	+1·2	+0·7	+0·5
CEUTA (Spanish enclave in Morocco)							
–0040	–0120	–0140	–0040	0·0	+0·1	+0·1	+0·1
ENSENADA DE TETOUAN (Morocco)							
–0045	–0045	No data		–0·1	0·0	+0·1	+0·1

SHELTER
Good in 3 marinas, from S to N: Queensway Quay, close to town centre; Sheppards at N end of town; and adjacent Marina Bay, close to airport runway. Advisable to pre-book. All can be affected by swell in W'lies and by fierce gusts in the E'ly *levanter*. ⌓ in 4m, to NW of runway (well clear of flight path); pontoons are for small local craft only.

NAVIGATION
WPT 36°06'·00N 05°23'·00W, 250°/070° from/to Europa Point, 2M. The Rock is steep-to all round; from the SW, beware shoals to S of Pta Carnero. From the E, yachts may round Europa Pt 3ca off. In the Bay commercial vessels at ⌓ and navigational lights may be hard to see at night. Note: Yacht arrivals for Sheppards and Marina Bay **must** clear in at adjacent conspic Customs berth (Waterport); Queensway Quay will clear yachts by fax. Visiting RIBs (Rigid Inflatable Boats) must obtain prior written permission to enter from Collector of Customs, Customs House, Waterport, Gibraltar, ☎ 78362; entry may otherwise be refused.

LIGHTS AND MARKS
The Rock (423m) is not easily missed, but from the W is not seen until rounding Pta Carnero, 5M to SW. On top of Rock is powerful Aero lt, Mo (GB) R 10s 405m 30M. Europa Pt, less obvious, has 3 sectored lts; see IG 27. At N end of bay, industrial plants are conspic by day/night.

RADIO TELEPHONE
Queensway Quay and Marina Bay VHF Ch 73; Sheppards Ch 71. Civil port Ch 06; Gibraltar Bay Ch 12; QHM Ch 08.

TELEPHONE
(International code 350, no Area code); from Spain dial 956 7 + 5 digit tel no).
Port Captain (Hr Mr) 77254; QHM via HM Naval Base; Marinas, see below; ⌗ 78879, ☎ 78362; Met 53416; ⊞ 79700; Ambulance/Police 199; Fire 190; ✈ 75984.

FACILITIES
Queensway Quay Marina, PO Box 19. (120+ Ⓥ), ☎ 44700, ☎ 44699, AB/F&A £6.50 in 3·5m least depth, AC, FW, CH, D & P (min 200ltrs from tanker) by arrangement, ▣;
Sheppards Marina, Waterport, Gib. (150, inc Ⓥ on F & G pontoons), ☎ 75148, ☎ 42535, AB/F&A £6, AC, FW, BH (40 ton), C (10 ton), ME, El, BY, Sh, CH, Ⓔ;
Marina Bay Marina, PO Box 80. (209 inc Ⓥ) ☎ 73300, ☎ 78373, AB £6.25 (max draft 4·5m), AC, FW, CH, R, V, ▣;
Services: ACA, SM; D & P, Gaz, Ice from Fuel berth (near Customs); **Royal Gibraltar YC. Town**: V, R, Bar, ✉, Ⓗ, Ⓑ, ✈.

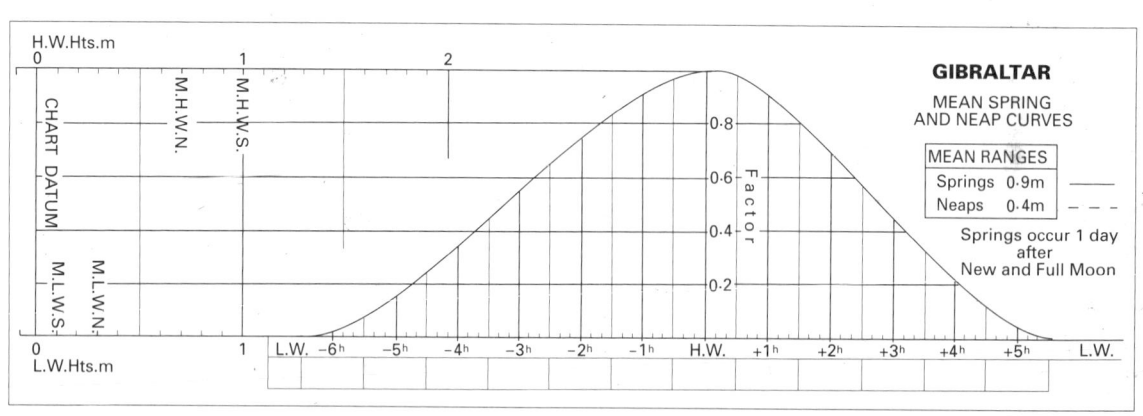

GIBRALTAR
MEAN SPRING AND NEAP CURVES

MEAN RANGES	
Springs	0·9m
Neaps	0·4m

Springs occur 1 day after New and Full Moon